Introduction to Anatomy

INTRODUCTION TO ANATOMY

Edited by:

Krunal Bhatt and Dipak Gunvanta Ingole

Delve PUBLISHING

www.delvepublishing.com

Introduction to Anatomy

Krunal Bhatt and Dipak Gunvanta Ingole

Delve Publishing
224 Shoreacres Road
Burlington, ON L7L 2H2
Canada
www.delvepublishing.com
Email: orders@arclereducation.com

ISBN: 978-1-77469-511-1 (Hardcover)

ABOUT THE EDITORS

Dr Krunal Bhatt (1982) is presently working at Zydus hospital, Vadodara as a chief clinical intensivist and Anaesthesiologist for one year. He did his MBBS from Baroda Medical College. He has worked as a Medical Officer at various corporate hospitals in Vadodara (2006-2009). He joined Indian armed forces and posted to various locations during his tenure (2009-2018). He has served at Pune, Lucknow, Jodhpur, Jammu and Kashmir. He also has obtained Counter Insurgency ops medal and High Altitude serving medals for his serving tenure at line of control in Kashmir. He did his post-graduation in Anaesthesiology from AFMC, Pune, India in 2018. He is Life Member of Indian Medical Association, Indian Society of Anesthesiology, Indian Society of Critical Care Medicine, and European Society of Anaesthesia.

Dr. Dipak Ingole (1989) is presently serving as Assistant Professor, Department of Genetics and Plant Breeding, CSMSS, College of Agriculture, Kanchanwadi, Aurangabad (MH). He obtained his B. Sc. (Ag.) in 2011 from Dr. Panjabrao Deshmukh Krishi Vidyapeeth, Akola, M. Sc. (Ag.) in 2011 and Ph. D. in Genetics and Plant Breeding in 2017 from Vasantrao Naik Marathwada Krishi Vidhyapeeth, Parbhani. He started his career as Assistant professor in 2017. He has published 02 books and 11 research paper in National and international reputed journals. He has participated in many National, State level Seminars and symposiums. He has delivered several Radio talks, Lectures in farmer training programme. Dr. Dipak has also written 35 popular articles in reputed Magazines. He has also received Young scientist award in 2022.

TABLE OF CONTENTS

LIST OF FIGURES

LIST OF ABBREVIATIONS

3D	Three Dimensional
ABA	Abscisic Acid
ABP	Androgen Binding Protein
AC-PC	Anterior Commissure-Posterior Commissure
AHK	Arabidopsis Histidine Kinase
AI	Artificial Intelligence
AMH	Antimullerian Hormone
An	Anthranilic Acid
AOC	Allene Oxide Cyclase
AR	Androgen Receptor
AR	Augmented Reality
ARRs	Arrabidopsis Response Regulators
ARs	Adventitious Roots
ART	Assisted Reproductive Technology
ASA1	Anthranilate Synthase A1
ATF6	Activating Transcription Factor 6
BG	Basal Ganglia
BiP	Binding Protein
BR	Brassinosteroid
BTB	Blood-Testis Barrier
CBME	Competency-Based Medical Education
CNS	Central Nervous System
CPT-1	Cytosolic Carnitine Palmitoyltransferase-1
CRF2	Cytokinin Response Factor 2
CT	Computed Tomography
DBS	Deep Brain Stimulation
DCD	Developmental and Cell Death
DDFT	Deep Digital Flexor Tendon
DHT	Dihydrotestosterone

DVT	Deep Vein Thrombosis
ECM	Extracellular Matrix
ER	Endoplasmic Reticulum
ERD	Early Dehydration Responsive
FAWC	Farm Animal Welfare Council
FDA	Food and Drug Administration
FSH	Follicle-Stimulating Hormone
GA	Gibberellic Acid
GABA	Gamma-Aminobutyric Acid
GLUT	Glucose Transporter
GNDF	Glial Cell Line-Derived Neurotrophic Factor
GnRH	Gonadotropic Releasing Hormone
H&E	Hematoxylin and Eosin
HCG	Human Chorionic Gonadotropin
HSPs	Heat shock proteins
IAA3	Indole-3-Acetic Acid 3
IBL	Internet-Based Learning
IFTS	International Federation of Teratology Societies
IRE1	Inositol-Requiring Enzyme 1
IVF	*In vitro* Treatment
JA	Jasmonic Acid
KEGG	Kyoto Encyclopedia of Genes and Genomes
LEA1	Late Embryogenesis Abundance 1
LH	Luteinizing Hormone
LMS	Learning Management System
LRs	Lateral Roots
LSPV	Left Superior Pulmonary Vein
LVOT	Left Ventricular Outflow Tract
MAKR4	Membrane-Associated Kinase Regulator 4
MCT	Medium-Chain Triglycerides
MER	Microelectrode Recording
MHW	Ministry of Health and Welfare
MIS	Mullerian-inhibiting Substance
MOOC	Massive open online courses

MR	Magnetic resonance
MRI	Magnetic Resonance Imaging
NMR	Nuclear Magnetic Resonance
PABP	PolyA-Binding Proteins
PAT	Polar Auxin Transport
PFO	Patent Foramen Ovule
PL1	Phospholipases1
PMC	Peritubular Myoid Cells
PR	Primary Root
RAM	Root Apical Meristem
RCVS	Royal College of Veterinary Surgeons
ROS	Receptive Oxygen Species
RSA	Root System Architecture
RSPV	Right Superior Pulmonary Vein
RV	Right Ventricle
RVOT	Right Ventricular Outflow Tract
SAM	Shoot Apical Meristem
SDFT	Superficial Digital Flexor Tendon
SHY2	Short Hypocotyl 2
SRY	Sex-Determining Region of The Y Chromosome
SSC	Spermatogonia Stem Cells
STN	Subthalamic Nucleus
SV	Soleus Veins
TDF	Testis Characterized Factor
TF	Transcription Factors
TIR1	Transport Inhibitor Response 1
UPDRS	Unified Parkinson Disease Rating Scale
VA	Virtual Anatomy
VHP	Visible Human Project
VLE	Virtual Learning Environment
VPE	Vacuolar Processing Enzyme
VR	Virtual reality

GLOSSARY

A

Adolescence - is the phase of life between childhood and adulthood, from ages 10 to 19.

Apoplasts - Inside a plant, the apoplast is the space outside the plasma membrane within which material can diffuse freely. It is interrupted by the Casparian strip in roots, by air spaces between plant cells and by the plant cuticle.

Apprenticeship - is a system for training a new generation of practitioners of a trade or profession with on-the-job training and often some accompanying study.

Arrhythmias - is an irregular heartbeat.

Auxin - Any of a group of plant hormones that regulate growth, particularly by stimulating cell elongation in stems.

B

Basal ganglia - are a group of subcortical nuclei, of varied origin, in the brains of vertebrates.

C

Cadaver - A cadaver is a dead body, especially a dead human body. The word cadaver is sometimes used interchangeably with the word corpse, but cadaver is especially used in a scientific context to refer to a body that is the subject of scientific study or medical use, such as one that will be dissected.

Catheter ablation - is a minimally invasive treatment for fast heartbeats.

Chromatography - Chromatography is a laboratory technique for the separation of a mixture into its components. The mixture is dissolved in a fluid solvent called the mobile phase, which carries it through a system on which a material called the stationary phase is fixed.

Cortical - relating to the outer layer of the cerebrum.

Cucurbita - Cucurbita is a genus of herbaceous vegetables in the gourd family, Cucurbitaceae native to the Andes and Mesoamerica. Five species are grown worldwide for their edible vegetable, variously known as squash, pumpkin, or gourd, depending on species, variety, and local parlance, and for their seeds.

Curricula - all the subjects that are taught in a school, college or university; the contents of a particular course of study.

Cytokinins - Cytokinins are a class of plant hormones that promote cell division, or cytokinesis, in plant roots and shoots. They are involved primarily in cell growth and differentiation, but also affect apical dominance, axillary bud growth, and leaf senescence.

D

Diagnosis - the act of saying exactly what illness a person has or what the cause of a problem is.

Differentiation - the process in which a stem cell alters from one type to a differentiated one.

Dissection - Dissection (from Latin dissecare "to cut to pieces"; also called anatomization) is the dismembering of the body of a deceased animal or plant to study its anatomical structure. Autopsy is used in pathology and forensic medicine to determine the cause of death in humans.

E

Electrophysiologists - are doctors who specialize in treating cardiac problems involving electrical activity and arrhythmia.

Elongated - to make (something) longer or to grow longer lengthen.

Ergastic - Ergastic substances are non-protoplasmic materials found in cells. The living protoplasm of a cell is sometimes called the bioplasm and distinct from the ergastic substances of the cell.

F

Fishbowl technique - is a strategy for organizing medium- to large-group discussions.

Foldases - In molecular biology, foldases are a particular kind of molecular chaperones that assist the non-covalent folding of proteins in an ATP-dependent manner. Examples of foldase systems are the GroEL/GroES and the DnaK/DnaJ/GrpE system.

G

Gladiator - A Roman gladiator was an ancient professional fighter who usually specialized in particular weapons and types of armor.

L

Light Microscope - The optical microscope, also referred to as a light microscope, is a type of microscope that commonly uses visible light and a system of lenses to generate magnified images of small objects.

Limb - An arm or leg of a person or four-legged animal, or a bird's wing.

M

Meristems - The meristem is a type of tissue found in plants. It consists of undifferentiated cells capable of cell division. Cells in the meristem can develop into all the other tissues and organs that occur in plants. These cells continue to divide until a time when they get differentiated and then lose the ability to divide.

Midbrain - is the topmost part of the brainstem, the connection central between the brain and the spinal cord.

Morphology – A branch of biology that deals with the form and structure of animals and plants · the form and structure of an organism or any of its parts.

N

Nneuropore- is a region corresponding to the opening of the embryonic neural tube in the anterior portion of the developing prosencephalon.

O

Organogenesis - Organogenesis is the phase of embryonic development that starts at the end of gastrulation and continues until birth. During organogenesis, the three germ layers formed from gastrulation form the internal organs of the organism.

P

Paleolithic age - Ancient technological or cultural stage characterized using rudimentary chipped stone tools.

Pedagogy - the method and practice of teaching, especially as an academic subject or theoretical concept.

Pericycle - The pericycle is a cylinder of parenchyma or sclerenchyma cells that lies just inside the endodermis and is the outer most part of the stele of plants.

Phenotype - refers to the observable physical properties of an organism; these include the organism's appearance, development, and behavior.

Physiologic - something that is normal, that is due neither to anything pathologic nor significant in terms of causing illness.

Physiology - Physiology is the study of how the human body works. It describes the chemistry and physics behind basic body functions, from how molecules behave in cells to how systems of organs work together. It helps us understand what happens in a healthy body in everyday life and what goes wrong when someone gets sick.

Pliable - easy to bend or shape.

R

Renaissance Period - The Renaissance was a fervent period of European cultural, artistic, political and economic "rebirth" following the Middle Ages. Generally described as taking place from the 14th century to the 17th century, the Renaissance promoted the rediscovery of classical philosophy, literature and art.

Senescence - Senescence is a process in which cells reach permanent growth arrest without the death of cells as the whole cell division process comes to a halt.

Septum - is the cartilage in the nose that separates the nostrils.

Spermatozoa - are the male sex cells that carry a man's genetic material.

T

Theophrastus - Theophrastus, a Greek native of Eresos in Lesbos, was the successor to Aristotle in the Peripatetic school. His given name was Tyrtamus; his nickname Θεόφραστος was given by Aristotle, his teacher, for his 'divine style of expression'. He came to Athens at a young age and initially studied in Plato's school.

Trabecular - is a highly porous (typically 75–95%) form of bone tissue that is organized into a network of interconnected rods and plates called trabeculae which surround pores that are filled with bone marrow.

Transcriptome - A transcriptome is the full range of messenger RNA, or mRNA, molecules expressed by an organism. The term "transcriptome" can also be used to describe the array of mRNA transcripts produced in a particular cell or tissue type.

V

Vasculature - the vascular system of a part of the body and its arrangement.

ABSTRACT

Anatomy is a branch of biology that is concerned with the identification and description of living organisms' internal structures. It comes from the Greek words ana and tomia, which mean up and cutting, respectively, and denotes cutting up or dissection when combined. The present book describes the historical aspects of anatomy, Anatomy of humans, Veterinary Anatomy and plants. This book also covered technological adoption in anatomy education. Use of virtual lab, three-Dimensional Digital Simulations, cloud technology, and Artificial intelligence in Anatomy education has also been covered. This book will be helpful for students, researchers and scholars to understand the complexity of anatomical structures of humans, animals and plants.

PREFACE

This book introduces the readers to the basic concept of anatomy. It sheds light on the overview, related history and different types of anatomy. In addition, this book also emphasizes on plant science and its anatomy along with the current ICT tools used in teaching as well as learning anatomy.

The first chapter gives the readers an overview of anatomy and its history. Types of anatomy and their impact on medicine have also been discussed towards the end of this chapter.

The second chapter takes the readers through the human anatomy. It will provide knowledge about gray's anatomy, the culture of dissection, the humanistic face of anatomy, and ethical and future dimensions of anatomy. Towards the end, this chapter talks about innovative technologies being used in medical education and the most important area of study-human brain anatomy.

After that, the third chapter explains the cardiac anatomy for the electrophysiologist that includes an overview of heart anatomy. Also, it talks about anatomical, biological, and surgical features of basal ganglia. Later, midbrain embryology and anatomy are discussed to give a clear insight to the reader.

The fourth chapter provides insights on male reproductive anatomy, including the fundamental component of the male reproductive system along with their functions. It also explains the positional relationships among male reproductive organs in insects. Epigenetics in male fertility has also been discussed later in the chapter.

Then the fifth chapter introduces the readers to plant science - structure, anatomy, and physiology which comprise phloem and cell types, parenchyma, anatomy and root development, formation of roots, etc. Also, the chapter deals with the concepts related to medicinal plants, their parts, uses and ecology.

The sixth chapter sheds light on veterinary anatomy and physiology. It talks about the history of veterinary anatomy and physiology, and different imaging technologies that are being used in anatomy and physiology. In addition, the chapter also discusses the anatomy, histology and physiology of the healthy and lame equine hoof and veterinarians' role in animal welfare and behavioral assessment towards the end.

In the last chapter of this book, current trends in teaching and learning anatomy are discussed, in which the beginnings of anatomy and the classic teaching model are explained, along with some basic models for teaching and learning human anatomy. Technological trends in anatomy education are also given. Towards the end of this chapter, some challenges and opportunities in anatomy teaching are explained.

This book has been designed to suit the knowledge and pursuit of the researcher and scholars and to empower them with various aspects of anatomy so that they are updated with the information. I hope that the readers find the book explanatory and insightful and that this book is referred by scholars across interdisciplinary fields.

CHAPTER 1

AN OVERVIEW OF ANATOMY

CONTENTS

Anatomy is known to be a field in the biological sciences which is concerned with the identification and description of the body structures of living things. Thus, it is important to have detailed knowledge of it, especially for science students. The chapter initially talks about the basic overview of anatomy and related history. Types of anatomy and its future have also been discussed. In the end, the impact of anatomy on medicine has been explained.

1.1 INTRODUCTION

Anatomy is a branch of biology that is concerned with the identification and description of living organisms' bodily structures. Gross anatomy is the study of major bodily structures through dissection and observation, and it is limited to the human body in its application.

While on the one hand, microscopic anatomy is concerned with the study of structural units tiny enough to be seen only with a light microscope, gross anatomy is concerned with the study of bodily components large enough to be studied without the need of magnifying devices. All anatomical research requires dissection.

The Greeks were the first to employ dissection, and Theophrastus dubbed it "anatomy," from *ana temnein*, which means "to cut apart." The field's other major subsection, comparative anatomy, evaluates similar body structures in different species of animals in order to comprehend the adaptive changes that have occurred through time. Anatomy is a field of biology concerned with the description of bodily structures revealed through dissection in various living organisms.

The word anatomy comes from the Greek word "anatom," which means "cutting" and "ana" means "up." Anatomy was first taught by dissecting corpses, hence the name "anatomy." When studying about the body and how it operates, quite frequently the phrase "anatomy" is coined. But what exactly is anatomy? What does anatomy mean, how would one define anatomy? Well, anatomy is a biological field of science that deals with the structure and identification of organisms' bodies and their many divisions.

Though the term "anatomy of the body" is commonly used in reference to humans and human body parts, it actually refers to all living organisms. Gross or macroscopic anatomy and microscopic anatomy are the two divisions of this study of body structure.

Between 1500 and 1850, this ancient discipline reached its pinnacle, with its subject matter solidly established. None of the world's ancient civilizations dissected a human body, which most people linked with the spirit of the departed soul and revered with superstitious awe.

Beliefs in life after death, as well as a nagging doubt about the likelihood of corporeal resurrection, further hampered rigorous research and systematic study. Nonetheless, treating wounds, assisting in childbirth, and setting broken limbs provided an understanding of the body.

However, until the successes of the Alexandrian medical school and its most famous figure, Herophilus (flourished 300 BCE), who dissected human cadavers and therefore provided anatomy a substantial factual basis for the first time, the field remained theoretical rather than descriptive. Herophilus produced numerous important discoveries, and his younger contemporary Erasistratus, who is frequently regarded as the founder of physiology, followed in his footsteps.

Galen, a Greek physician, gathered and organized all of the Greek anatomists' discoveries in the 2nd century CE, including his own notions of physiology and discoveries in experimental medicine. Because they were the only ancient Greek anatomical textbooks that survived the Dark Ages in the form of Arabic (and then Latin) translations, Galen's many books became the undisputed authority for anatomy and medicine in Europe.

Because of religious prohibitions against dissection, European medicine in the Middle Ages relied on Galen's combination of fact and fiction rather than direct observation for anatomical knowledge, while occasional dissections were permitted for educational purposes.

Leonardo da Vinci performed his own dissections in the early 16th century, and his beautiful and accurate anatomical drawings paved the way for Flemish physician Andreas Vesalius to "restore" the science of anatomy with his monumental De humani corporis fabrica libri septem (1543; "The Seven Books on the Structure of the Human Body"), the first comprehensive and illustrated textbook of anatomy.

As a professor at the University of Padua, Vesalius encouraged younger scientists to accept traditional anatomy only after independently verifying it, and this more critical and questioning attitude undermined Galen's authority, putting anatomy on a firm foundation of observed fact and demonstration. His Padua successors built on Vesalius' precise descriptions of the skeleton, muscles, blood vessels, nervous system, and digestive tract to study the digestive glands, as well as the urinary and reproductive systems.

Among the most significant Italian anatomists were Hieronymus Fabricius, Gabriello Fallopius, and Bartolomeo Eustachio, whose meticulous research led to fundamental advances in the allied subject of physiology. Fabricius' comprehensive descriptions of the venous valves, for example, were crucial in William Harvey's discovery of blood circulation.

Anatomy is a discipline of science concerned with the study of organisms' physical structures and parts. It comes from the Greek words ana and tomia, which mean up and cutting, respectively, and denotes cutting up or dissection when combined. Many consider it to be one of the earliest medical specialties.

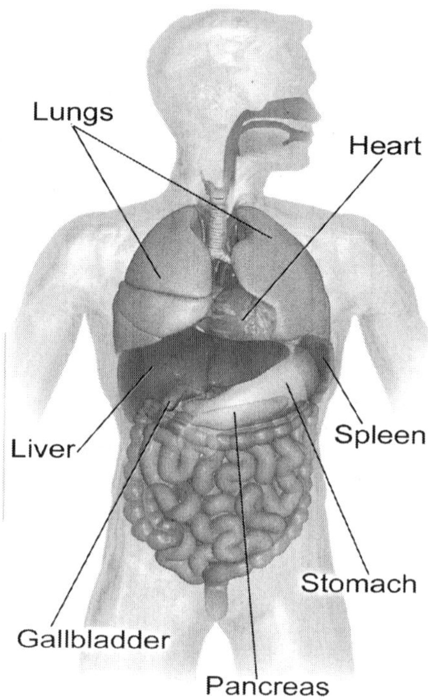

Figure 1.1. Abdominal organs anatomy.

Source: Image by Wikimedia commons

The development of anatomy was influenced by religious and philosophical issues. Animals such as pigs and monkeys were used in the early research before moving on to people. Alcmaeon, an ancient Greek scientist and philosopher who lived about the 5th century BCE, is thought to have been the first person to study anatomy.

He is also credited with performing the first human body dissection. His research led to a new understanding of the eyes, optic nerves, and the brain-eye link. Other well-known scientists like Hippocrates, Herophilus, and Aristotle would continue to improve on Alcmaeon's knowledge.

1.2 HISTORY OF ANATOMY

Humans are inquisitive organisms that have always been fascinated by how bodies are formed and created. Caves have been discovered with anatomical structures, hyroglifics, and drawings dating back to 750000 BCE. Despite the simplicity of the drawings of what was seen, these offered proof that the primordial human had a rather broad grasp of the anatomy of the human body.

It has also been revealed that people began to perform modest "medical" procedures as early as the Paleolithic Period. A small hole was drilled into the skull of a person who may have had mental problems. Despite the fact that the reasoning behind the operations was not necessarily scientific, some of the people who experienced them lived and demonstrated a breakthrough in this field of medicine.

Fast forward to the Ancient Romans, who learned more about anatomy by treating their injured gladiators. However, because dissection of human bodies was prohibited, this could only be done in secret. As a result, the Romans relied on animal anatomy and bodies to compare and contrast their discoveries with those of humans.

Galen, a researcher and experimentalist, became a practising physician around this time. He investigated macroscopic anatomy, the majority of which involved animal dissections. The discoveries he uncovered would have a long-term impact. Anatomy charts and sketches by eminent artists like Leonardo Da Vinci and Rembrandt van Rijn first appeared during the Renaissance period.

Some of the foundations for modern anatomical sketches and human body diagrams of the organ systems of the body were laid by these painters. After dissecting portions of human beings, the artists collaborated with scientists to create anatomically correct sketches. Andreas Vesalius, who contributed greatly to anatomy by his persistent study of dissected individuals and his work on the first-ever book on human anatomy, lived during this time period.

Figure 1.2. An anatomical dissection by Pieter Pauw in the Leiden anatom.

Source: Image by Wikimedia commons

From the 17th to the 20th century, a number of anatomists and physicians contributed to the development of the anatomy that is known today and also aided in the advancement of medical knowledge. Until that moment, scientists who wished to research human internal organs and systems using cadavers or dissecting living persons were discriminated against.

Abraham Flexner's famous article from the twentieth century emphasized the relevance of medical research using body organs and how dissections help with fundamental medical education and training. This aided in the normalization of anatomy and the advancement of medical discoveries.

The Beginnings - 3rd Century B.C.

Anatomy is the oldest medical science study. The first scientific dissections of the human body were performed in Alexandria as early as the third century B.C. Anatomists of the period studied anatomy by dissecting animals, particularly pigs and monkeys.

Claudius Galen (129-199) is the most famous physician in Ancient Greece, whose results are based only on animal studies and whose erroneous beliefs on human anatomy dominate and influence medical knowledge for nearly 1,000 years, until the Renaissance.

Although the Church does not explicitly prohibit anatomy, social authorities did not allow the dissection of human bodies until the 12th and even 13th centuries. This is why anatomical research has reached a stalemate. During the 13th and 14th centuries, there is a shift in the way anatomy is taught. Teaching, on the other hand, consists primarily of lectures based on Galen's canonical books, with no actual dissections.

Modern Age - 15th/16th Century

Many anatomical dissections of human bodies are performed by Leonardo da Vinci (1452-1519). In fact, today's most well-known Renaissance artist and scientist, and serve as the foundation for his famous, extremely detailed anatomical illustrations.

Anatomy & Art Leonardo Da Vinci

The body was viewed as the frail home of the spirit in medieval times. The human body, on the other hand, is elevated for its beauty throughout the Renaissance, and it becomes the primary source of inspiration for painters of the time. Many Renaissance artists begin researching the human body for the sake of art.

Leonardo da Vinci and Michelangelo did not simply watch their medically educated acquaintances do dissections; they take up the scalpel themselves in order to depict the body in all its natural grandeur. The bone structure, skeleton, and skin are all shown in the most realistic way possible, in addition to the body and muscles.

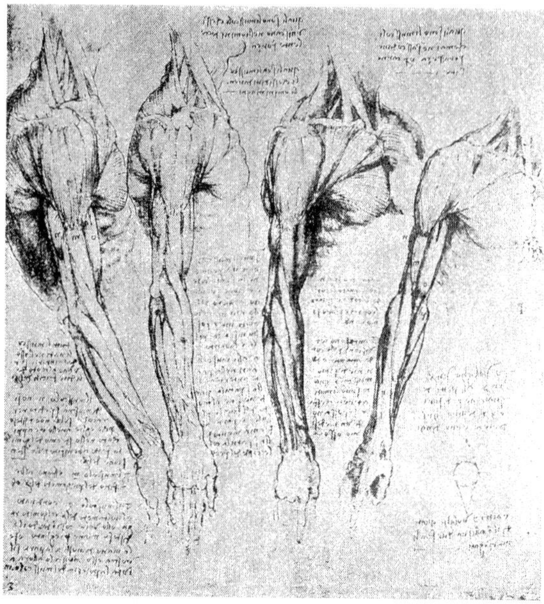

Figure 1.3. Leonardo da Vinci the anatomist from McMurrich.

Source: Image by Wikimedia commons

Leonardo da Vinci was a man who was fascinated by the human body. He used to climb cemetery walls in the dead of night, snatch remains, and drag them into his workshop where he would dissect them and use them as models for his sculptures.

From The 16th Century Onwards

Andreas Vesalius, an anatomist and surgeon, developed the actual science of anatomy during the Renaissance. During the public dissection of human bodies, Vesalius reports what he observed. Vesalius was able to establish more than 200 inaccuracies in Galen's anatomical writings by dissecting human bodies and meticulously preparing muscles, tendons, and nerves.

The young professor of medicine revolutionized anatomy and, as a result, the entire discipline of medicine, with his detailed scientific examinations of human bodies. During the Renaissance, the dissections were not only of interest to a medical forum but also accessed by the broader public.

The frontispiece illustration for Andreas Vesalius' 7-volume opus, "On the Fabric of the Human Body" made it quite evident too. Essentially it showed Vesalius performing a dissection in a crowded theatre.

Anatomical Theaters - 17th Century

The Renaissance anatomists were inspired by artistic passion, and interest in anatomy grows among the masses. Physicians, as well as the general public, had started becoming increasingly interested in seeing the human body firsthand.

In fact, "To see with one's own eyes" was a Greek term that inspired the word "autopsy."Moreover. many cities erected anatomical theatres, and the public dissection presentations would draw people from all walks of life.

Anatomical Art - 18th Century

Some anatomists employed their dissecting talents in a more traditional artistic manner, transforming their specimens into permanent pieces of art. Honoré Fragonard transforms his anatomical specimens into works of art that will last a lifetime. He would inject colored wax into their blood vessels, which used to harden up.

After this, the leftover tissues were allowed to dry before being varnished. His work is still on exhibit at the Ecole Nationale Vétérinaire d'Alfort in Alfort, France, just outside of Paris. Anatomical painters made the first whole-body specimens in the 18th century, which were dried and varnished. Metal alloys were then melted and injected into the arteries, while still hot, in some specimens from that time.

Modern Anatomy - 19th/20th Century

After establishing the basics of human macroscopic anatomy (the study of dissected organs), the subject of anatomy became increasingly specialized, and anatomical knowledge expanded into the microscopic level. For ages, the public's fascination with anatomy did not wane. In fact, the public was not allowed to see dissections until the nineteenth century, when anatomy becomes a science.

1.3 TYPES OF ANATOMY

Gross anatomy, microscopic anatomy, human anatomy, phytotomy, zootomy, embryology, and comparative anatomy are some of the various types of anatomies.

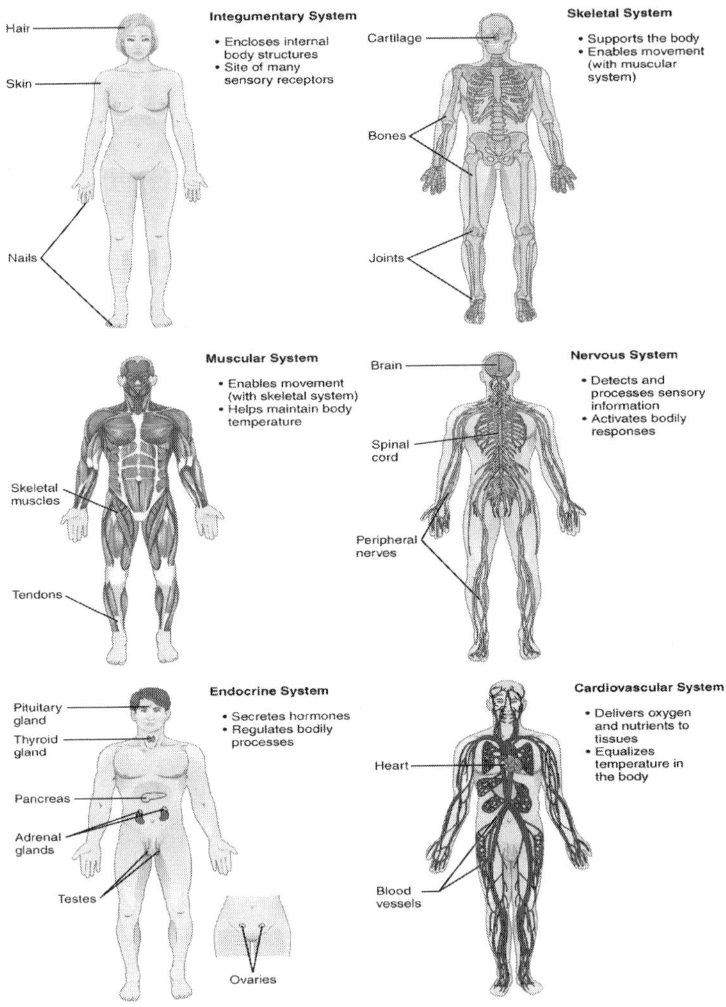

Figure 1.4. Organ systems of the body.

Source: Image by Wikimedia commons

Gross Anatomy

The macroscopic or large-scale examination of organs and structures as a whole is known as gross anatomy. It's also known as the study of organs and structures that can be seen with the naked eye. Gross anatomy is one of the branches of anatomy. On a macroscopic level, this is the study of anatomy that can be seen with the naked eye.

Cytology and histology are enhanced by gross anatomy. Histology is the study of tissues and their architecture, whereas cytology is the study of cells as the initial basic unit of life. Because gross anatomy details the development of each attribute of an organism's body part, it frequently goes hand in hand with growth and development.

Between 1500 and 1850, this ancient discipline reached its pinnacle, with its subject matter firmly established. None of the world's ancient civilizations dissected a human body, which most people linked with the spirit of the departed soul and revered with superstitious awe. Beliefs in life after death, as well as a nagging doubt about the likelihood of corporeal resurrection, further hampered rigorous research.

Nonetheless, treating wounds, assisting in childbirth, and setting broken limbs provided an understanding of the body. However, until the successes of the Alexandrian medical school and its most famous figure, Herophilus (in 300 BCE), who dissected human cadavers and therefore provided anatomy a substantial factual basis for the first time, the field remained theoretical rather than descriptive.

Herophilus produced numerous important discoveries, and his younger contemporary Erasistratus, who is frequently regarded as the founder of physiology, followed in his footsteps. Galen, a Greek physician, gathered and organized all of the Greek anatomists' discoveries in the 2nd century CE, including his own notions of physiology and discoveries in experimental medicine.

Because they were the only ancient Greek anatomical textbooks that survived the Dark Ages in the form of Arabic (and then Latin) translations, Galen's many books became the undisputed authority for anatomy and medicine in Europe. Due to religious prohibitions against dissection, European medicine in the Middle Ages relied on Galen's combination of fact and fiction rather than direct observation for anatomical knowledge, while occasional dissections were permitted for educational purposes.

Leonardo da Vinci performed his own dissections in the early 16th century, and his beautiful and accurate anatomical drawings paved the way for Flemish physician Andreas Vesalius to "restore" the science of anatomy with his monumental De humani corporis fabrica libri septem (1543; "The Seven Books on the Structure of the Human Body"), the first comprehensive and illustrated textbook of anatomy.

As a professor at the University of Padua, Vesalius encouraged younger scientists to accept traditional anatomy only after independently verifying it, and this more critical and questioning attitude undermined Galen's authority, putting anatomy on a firm foundation of observed fact and demonstration.

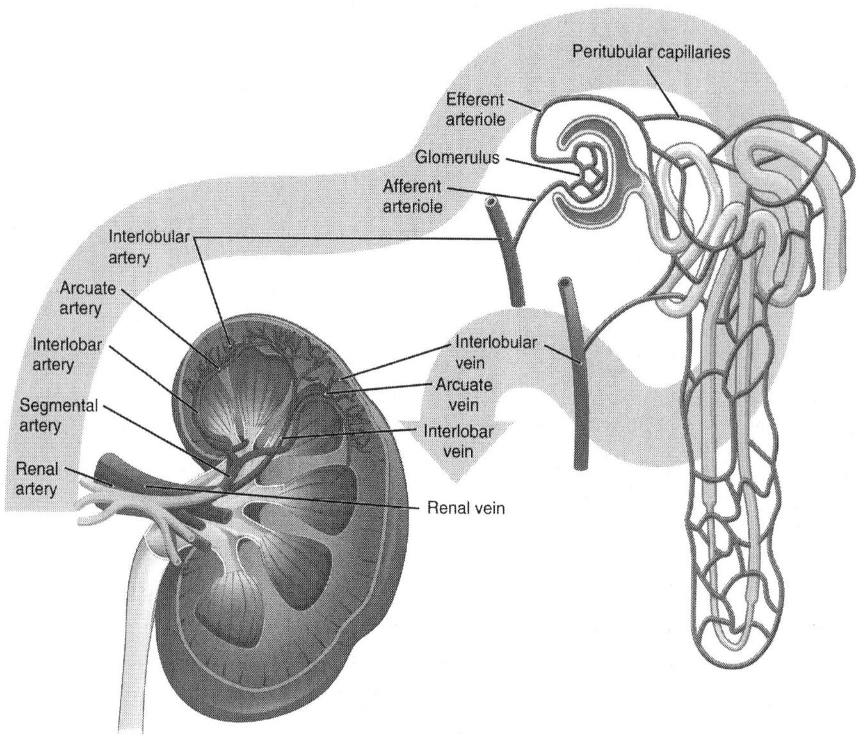

Figure 1.5. Blood flow in the kidneys.

Source: Image by Wikimedia commons

Vesalius's Padua successors built on his precise descriptions of the skeleton, muscles, blood vessels, nervous system, and digestive tract to study the digestive glands, as well as the urinary and reproductive systems. Among the most significant Italian anatomists were Hieronymus Fabricius, Gabriello Fallopius, and Bartolomeo Eustachio, whose meticulous research led to fundamental advances in the allied subject of physiology.

Fabricius' comprehensive descriptions of the venous valves, for example, were crucial in William Harvey's discovery of blood circulation. Surface, regional, and systemic anatomy are subsets of macroscopic anatomy. These are concerned with the outside of the body, regions of the body, and specialized body systems, respectively.

While one is not required to dissect anything to understand and observe surface anatomy, this division examines the exterior body form and what it accomplishes to allow the body to function while safeguarding interior systems.

Regional anatomy examines specific parts of the body and how they interact to perform a variety of tasks. When dealing with systemic anatomy, the gastrointestinal system and the circulatory system are two examples of the eleven (11) body systems that are focused on.

Microscopic Anatomy

Microscopic anatomy, as the name suggests, is the study of the structure of organisms at a microscopic level. Microscopic anatomy and histology are often used interchangeably, although this is erroneous because microscopic anatomy encompasses both histology and cytology. Histology is the study of how cells, which are the body's building blocks, evolve from cells to tissues to organs and organ systems. And these several stages of development come together to form a live creature.

Microscopic anatomy, on the other hand, is limited to tissues and smaller entities since they are the only ones that fit under the microscope. Microscopic anatomy includes both histology and cytology, as previously stated. Both of these procedures necessitate the thin slicing of organs in order to produce specimens for microscopy. This can be done on both alive and dead cells and tissues.

Furthermore, in order to obtain a contrast and visibility between the organelles and components of the tissues and cells, they are dyed. This method makes studying the anatomy of miniature body parts easier. The most essential aspect in the future growth of anatomical research was the innovative use of magnifying glasses and compound microscopes in biological studies in the second half of the 17th century.

Marcello Malpighi discovered the system of microscopic capillaries connecting the artery and venous networks, Robert Hooke observed the first minute compartments in plants that he dubbed "cells," and Antonie van Leeuwenhoek observed muscle fibers and spermatozoa using primitive early microscopes. From then on, the focus moved from identifying and comprehending physiological structures apparent to the naked eye to microscopic structures.

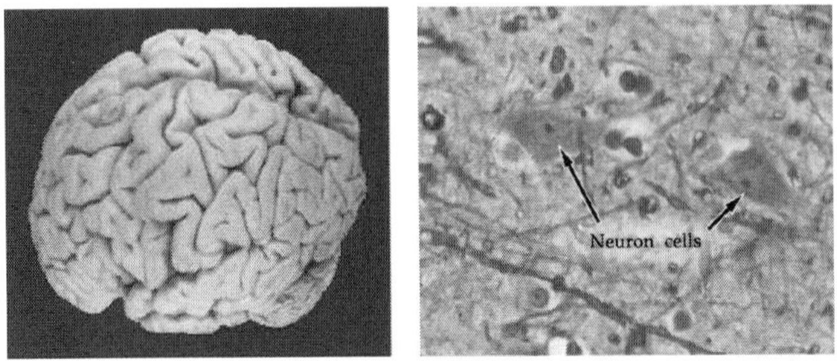

(a) (b)

Figure 1.6. Gross and Microscopic anatomy.

Source: Image by Wikimedia commons

In the 18th century, the use of the microscope in discovering minute, previously unknown features became more systematic, but progress was slow until technical improvements in the compound microscope itself, beginning in the 1830s with the gradual development of achromatic lenses, greatly increased that instrument's resolving power.

The cell was recognized as the essential unit of organization in all living things by Matthias Jakob Schleiden and Theodor Schwann in 1838–39, thanks to these technological developments. The need for thinner, more transparent tissue specimens for analysis under the light microscope prompted the invention of new dissection techniques, including microtomes, which slice specimens into incredibly thin slices.

In order to better distinguish the detail in these sections, synthetic dyes were used to stain tissues with different colors. By the late 1800s, thin slices and staining became a routine technique for microscopic anatomists. The fields of cytology (the study of cells) and histology (the study of tissue organization from the cellular level up) both evolved in the nineteenth century on the basis of microscopic anatomy data and tools.

Anatomists tended to investigate tinier and tinier units of the body structure in the 20th century as new technologies allowed them to discern features well beyond the resolution of light microscopes. These breakthroughs were made possible by the electron microscope, which, starting in the 1950s, sparked a flood of research into subcellular structures and quickly became the most important tool in anatomical research.

Around the same time, the use of X-ray diffraction to study the architecture of many different types of molecules found in living things spawned the new field of molecular anatomy.

Human Anatomy

For obvious reasons, this is the branch of anatomy that is most popular around the world. As humans, people are always interested in learning more about how the bodies work and function. However, one must first understand the structure or form of the body before one can understand how the sections of the body work. This is where the study of human anatomy comes into play.

The human body's organization is a unique system in and of itself. It begins with the cell, which is the basic unit of the human body. Later, these cells come together to form tissues which further form organs that combine to form organ systems. These 11 body systems then form the human body.

Every other body part is made up of cells, which are the smallest unit in the body. The stem cell is the most basic type of cell, capable of developing into any sort of specialized cell required by the organism. The human body has about 200 different types of specialized cells, ranging from reproductive cells that aid in the development of babies to red blood cells that transport oxygenated and deoxygenated blood throughout the body.

Tissues are collections of cells with similar functions that come together to act as a unit. Epithelial, muscular, connective, and nerve tissues are the four primary types of tissues. The major tissues of the glands, epithelial tissues, serve a variety of tasks, including absorption, secretion, and filtration. Connective tissues connect structures together, forming a strong support system and providing disease protection.

Muscle tissues are tissues that are specifically intended to aid in the movement of the body. These include the cardiac tissues that allow blood to be circulated around the body by contracting and relaxing the heart. Nervous tissues make up the brain, spinal cord, and nerves. These are in charge of keeping the body and its functions in check, including coordination.

Organs are formed when numerous different types of cells and tissues join together. Organs are structures that carry out certain tasks. The lung's job is to help blood cells receive oxygen and expel carbon dioxide by facilitating gas exchange. The skin protects and controls temperature, as well as acts as a barrier against invaders. Organ systems are formed when these organs operate together with other organs.

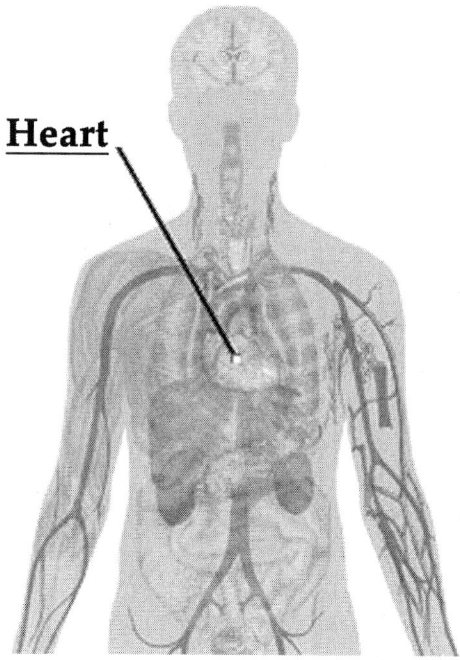

Figure 1.7. Human body diagram.

Source: Image by Wikimedia commons

There are eleven (11) organ systems in total. The following are examples of these systems:

- Cardiovascular System (Circulatory System)
- Digestive System (GI Tract)
- Endocrine System
- Integumentary System
- Lymphatic System
- Muscular System
- Nervous System
- Skeletal System
- Reproductive System
- Respiratory System
- Urinary System

Each of these contributes to the anatomy and physiology of the body in its unique way. The immune system is a biological system that overlaps with another system, the lymphatic system in particular.

Anatomical Nomenclature

The scientific names for the components and structures of the human body are often in Latin; for example, the biceps muscle of the upper arm is known as musculus biceps brachii. Many of these names were given to Europe by ancient Greek and Roman writers, and many more were invented by European anatomists beginning in the 16th century.

Many new bodily structures and tissues were discovered as medical knowledge grew, but there was no uniformity of nomenclature, and thousands of new names were created as medical writers followed their own whims, generally in Latin form. The confusion generated by the vast variety of names had grown untenable by the end of the nineteenth century.

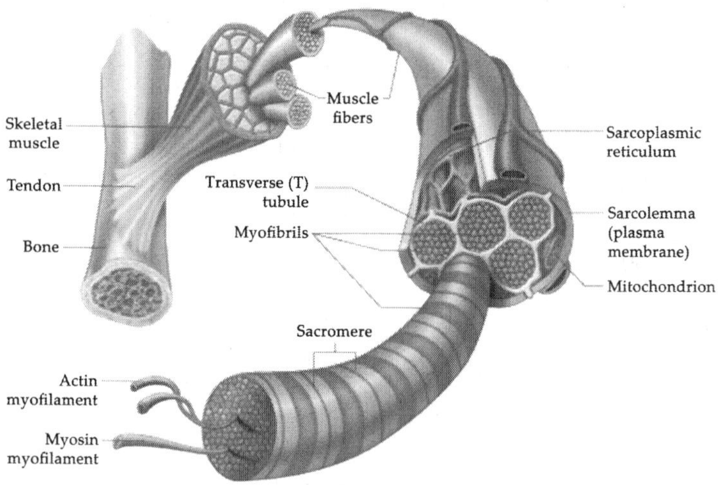

Figure 1.8. Muscular levels of organization.

Source: Image by pressbooks.ccconline.org

There were up to 20 synonyms for one name in medical dictionaries, and over 50,000 names were in use across Europe. The German Anatomical Society began the effort of standardizing nomenclature in 1887, and in 1895, a full list of anatomical words and names was accepted with the support of other national anatomical societies, reducing the 50,000 names to 5,528.

The Basle Nomina Anatomica had to be expanded later, and the Sixth International Anatomical Congress in Paris in 1955 accepted a substantial revision known as the Paris Nomina Anatomica (or simply Nomina Anatomica). The Terminologia Anatomica, which recognizes about 7,500 names defining macroscopic parts of human anatomy and is regarded as the international standard on human anatomical nomenclature, succeeded this book in 1998. In 2011, the International Federation of Associations of Anatomists and the Federative Committee on Anatomical Terminology (after known as the Federative International Programme on Anatomical Terminologies) published the Terminologia Anatomica, which was made accessible online.

Zootomy

The anatomical of animals is zootomy. This field of medicine is crucial for veterinary students to learn about. It is concerned with the study of an animal's interior structure, including its cells, tissues, organs, bones, and other organs. People have long been captivated by animal anatomy, with mural paintings showing animal superficial anatomy dating back to the Paleo-lithic age.

However, evidence suggests that scientific anatomical study may have first appeared in ancient Babylonia, albeit the tablets on which this was recorded have since disappeared, and the ruins show that Babylonian knowledge was in fact restricted. The most modern, up-to-date, complete, concise, and digital database of the microscopic anatomy of animals is the Microscopic Anatomy of Animals.

Its goal is to become the primary resource for animal microscopic anatomy, histology, and cell biology (invertebrates and vertebrates). The database is based on correlative morphology, which means that data is always applied to all levels of the animal, from gross anatomy to organs, tissues, cells, and ultrastructure.

The extensive crosslinking of data helps comparison analyses and data mining, allowing morphologists and researchers from various areas of biology to obtain morphological data, even from mysterious species.

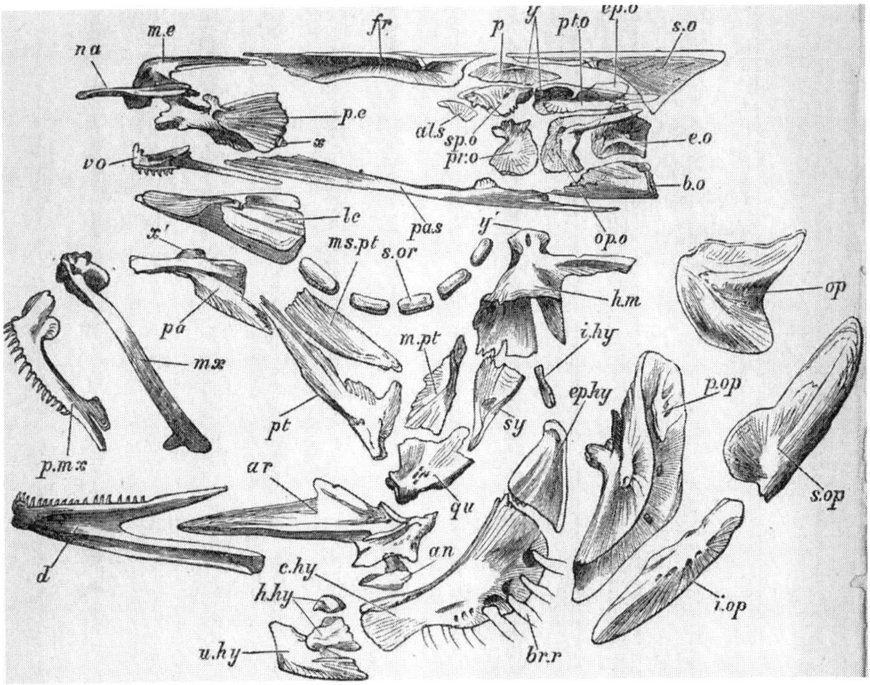

Figure 1.9. Zootomy.

Source: Image by Flickr

Phytotomy

Plant anatomy is the study of a plant's form, structure, and size. Plant anatomy is a branch of botany (the study of plants) that studies the structural elements and systems that make up a plant. A typical plant's body is made up of three basic vegetative organs: the root, stem, and leaf, as well as a set of reproductive organs including flowers, fruits, and seeds.

All of a plant's parts are made up of cells because it is a living thing. Plant cells, like animal cells, have a flexible membrane, but they also have a strong cellulose wall that gives them a solid form. Plant cells, unlike animal cells, contain chloroplasts, which absorb the Sun's light energy and convert it to food.

A plant, like any other complex living organism, organizes a collection of specialized cells into tissues that perform a specific function. Plants, for example, contain epidermal tissue that creates a protective coating on their surfaces. They also have parenchyma tissue, which is a type of energy storage tissue.

A plant's "veins," or pipelines, are made up of vascular tissue that transports water, minerals, and nutrients throughout the entire plant. Organs are formed when tissues are combined to fulfill a more sophisticated purpose. The roots of a plant, like the foundation of a skyscraper, assist it in remaining upright.

They also absorb water and dissolved minerals from the earth, providing the necessary ingredients for the plant to produce its own sustenance. Although the roots of some aquatic plants float, the majority of roots grow underground and travel downhill due to gravity. Other root systems, such as those of English ivy, actually adhere to vertical surfaces and allow the plant to climb.

Taproot and fibrous root systems are the two basic forms of root systems. Taproot plants have a single, lengthy root that penetrates straight down and anchors the plant firmly. Taproots are found in trees and dandelions, and they perform this purpose. Fibrous roots have a branching network and are shorter and shallower.

Grass has a fibrous root structure that develops in all directions at a shallow level. Pipelines or veins run through a root, carrying water and minerals to the rest of the plant. Like with the lead in a pencil, these pipes are concentrated in the root's core. Each root has a cap at the end that protects it as it penetrates deeper into the earth.

Continuing root hairs are found on the sides of the root, but further away from the root cap. These hairs are the plant's primary water and oxygen absorbers. Diffusion and osmosis are the two basic methods by which materials enter and leave roots. When molecules are scattered unevenly, nature constantly seeks a balance, and molecules will shift from a high to a low concentration location.

When the cells of a root hair have limited oxygen and the soil around the root hair has plenty, oxygen will naturally travel from the soil to the root without the plant having to exert any effort. Osmosis is a similar condition (from high to low concentration), but it happens when water molecules flow over a membrane that prevents other things from passing through. Osmosis, like diffusion, does not need the plant to utilize any energy.

Plant stems have two purposes. They sustain the plant's aboveground portions (typically the buds, leaves, and flowers), as well as transport water and nourishment within the plant. The epidermis is the outer layer of the stem; the cortex is the inner layer; and the pith is the core zone. Thousands of cells line up next to and on top of each other to support the stem of a green plant.

The cells expand like a full balloon as they take in water, and because their walls are elastic, they stretch very tightly against one other and the stem wall. The stem is held up by their pressure. When a plant's cells run out of water and begin to shrink, it droops. Lignin is a substance found in woody plants, such as trees, which strengthens and rigidifies cell walls.

The stem of a plant also serves as its circulatory system, with vascular tissue forming lengthy tubes through which materials are transported from the roots to the leaves and from the leaves to the roots.

Comparative Anatomy

Comparative anatomy is the study of distinct animal species' body architecture in order to understand the adaptive changes that have occurred over time as they evolved from common ancestors. The research of French naturalist Pierre Belon, who demonstrated in 1555 that the bones of humans and birds are made up of comparable parts organized in the same way, is the foundation of modern comparative anatomy.

With the research of two French naturalists—Georges-Louis Leclerc, comte de Buffon, and Louis-Jean-Marie Daubenton—who studied the anatomies of a wide range of species in the 18th century, comparative anatomy advanced significantly.

In the early nineteenth century, French biologist Georges Cuvier established a stronger scientific foundation for the topic by claiming that animals' structural and functional traits are the result of their interaction with their environment. Cuvier also disproved the 18th-century idea that the animal kingdom is organized in a single linear succession from the simplest to the most complex.

Instead, Cuvier classified all animals into four main classes based on their body plans (vertebrates, mollusks, articulates, and radiates). Another notable player in the discipline was mid-nineteenth-century British anatomist Sir Richard Owen, whose enormous understanding of vertebrate structure did not prevent him from criticizing British biologist Charles Darwin's idea of evolution by natural selection.

Comparative anatomy was used extensively by Darwin to further his theory, and it transformed the science by explaining structural differences between species as coming from their evolutionary descent from a common ancestor through natural selection. Since Darwin's time, comparative anatomy has mostly focused on homologous bodily structures, or those in distinct species that have the same evolutionary origin regardless of their current function.

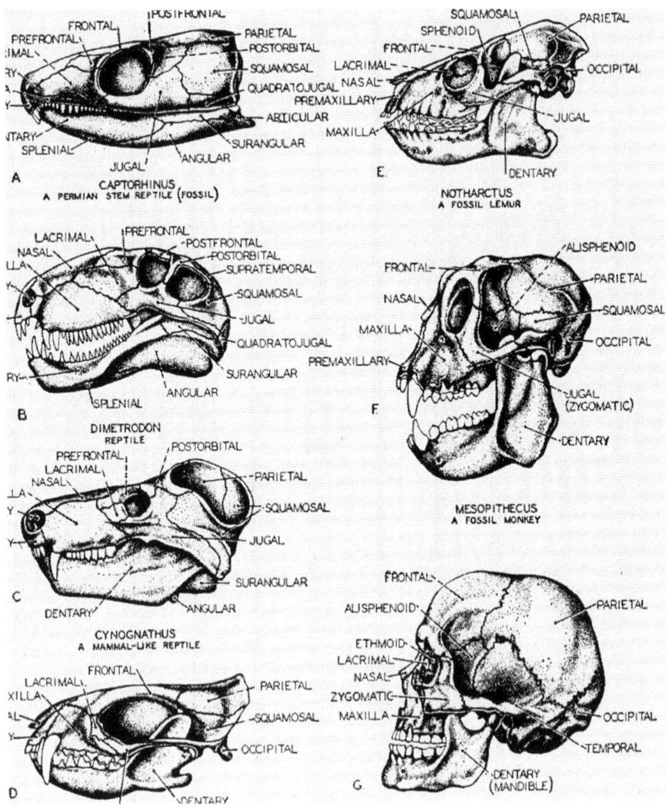

Figure 1.10. Comparative anatomy.

Source: Image by Flickr

4Although such structures may appear to be extremely distinct and have quite different functions, they may all be traced back to a common ancestor in an animal. Humans, birds, crocodiles, bats, dolphins, and rodents, for example, have evolved forelimbs that fulfill varied purposes, but they can all be traced back to the fins of crossopterygian fishes, where that basic arrangement of bones was initially developed.

Analogous structures, on the other hand, may look similar because they perform the same function, but they have different evolutionary origins and, in many cases, different structures, such as the wings of insects and birds.

1.4 ANATOMY: PAST, PRESENT AND FUTURE

The term "anatomy" comes from the Greek word "anatomy," which means "dissection," and refers to the research of the structure of organisms and their parts. Scientific minds have always been drawn to the creation of realistic anatomical representations in order to better depict and comprehend human anatomy, physiology, and illness.

The study of sacrificed human bodies predates the history of anatomy by approximately 1000 years. Greek scientists such as Alcmaeon (about 500 B.C.) and Hippocrates pioneered systematic anatomical study (ca. 460- 377 B.C.). Aristotle (approx. 384-322 B.C.) and his contemporaries developed a more descriptive anatomical system based on animal dissection in the following decades.

Herophilus (335–280 B.C.) is commonly regarded as the father of anatomy for performing the first systematic dissection of the human body. Galen (about 130-200 B.C.) dissected animals and authored anatomical treatises. The knowledge of Greek anatomical treatises was lost during the next few years; however, some were translated into other languages such as old English, Latin, and Arabic, and others were preserved in Byzantium and the Islamic world.

Medical science, anatomy, and medical education gradually regained prominence. Mondino de Luzzi (ca. 1270 – 1326) built the first European medical school in 1235 in Italy, including a methodical study of anatomy and dissection into a medical curriculum. The major mediums for representing anatomy were hand sketching on paper and woodcut engraving.

Artists participated in scientific depictions of the human body in the fifteenth century, notably Leonardo da Vinci's famous anatomical paintings (1452-1519). In his De Humani Corporis Fabrica, Andreas Vesalius (1514-1564), the father of modern human anatomy, added a new dimension to the subject by combining methodical text and illustration (On the Structure of the Human Body).

The contributions of scientists like Realdo Colombo (1515 – 1559) helped the discipline of anatomy reach new heights throughout the Renaissance. In his book "De Re Anatomica," Realdo Colombo described and called the placenta. In his book "Exercitatio Anatomica de Motu Cordis et Sanguinis in Animalibus," William Harvey (1578–1657) described how the heart drove blood in a circular path through the body.

New discoveries were produced, and experimental research methods were introduced to scientific inquiry. Microscopic anatomy was created by Marcello Malpighi (1628–1694). The speed and accuracy of manufacturing anatomical representations were increased because of new artistic processes such as copperplate engraving, copper plate etching, and lithography.

Anatomical drawings were done by many prominent painters during the Renaissance, including Michelangelo (1475 –1564) and Rembrandt (1606-1669). The invention of the printing press in the 15th and 16th centuries made these drawings more widely available through mass production. The fast expansion of medical schools in the sixteenth century creates a strong demand for cadavers.

Anatomical drawings become essential in the absence of cadavers. By the end of the 17th century, Gaetano Zumbo (1656–1701) had perfected anatomical wax modeling techniques, resulting in the first method of 3D anatomical modeling. Drying and wax injection were becoming more popular methods for storing and preserving dissected specimens.

Anatomical research was eventually expanded to include not just histology but also developmental biology in humans and animals. Wax embedding and histology could be conducted using Gaetano Zumbo's procedures. Scientists such as Giambattista Morgagni (1682-1771), Scott Matthew Baillie, and Xavier Bichat demonstrated the importance of "pathological anatomy" from the late 17th to early 19th centuries.

Scientists required a non-invasive approach to photograph the human body in order to match their anatomy expertise with disease physiology. Anatomical science has progressed significantly during the last hundred years. The way anatomy is taught has evolved as a result of new photographic mediums such as relief halftone and color photography.

Since Wilhelm Rontgen's invention of X-Ray machines in 1895, the inside of a human body has been visible without dissection. In 1930, Felix Bloch invented the Magnetic Resonance Imaging (MRI) system, which uses magnetic fields and radio waves to create images of organs by causing atoms to emit small radio signals.

Godfrey Hounsfield and Allan Cormack invented computed tomography (CT) in 1972, allowing researchers to see patients' organs in unprecedented detail. An effort was made to integrate three-dimensional (3d) anatomical representation in anatomy, teaching utilizing clay and plastic models, similar to Gaetano Zumbo's wax modeling.

Studying evolution, development, anatomical functioning, physiology, pathological assessment, and regenerative medicine are all important aspects of anatomy today. Although we currently have libraries of precise macroscopic and microscopic reproductions of human anatomy, current research orientations demand higher resolution.

Cadaveric body donations have decreased dramatically, owing to religious and disease transmission concerns, as well as some legislative statutes (such as the Human Tissue Act of 2004 in the United Kingdom), which limit the availability of human bodies to anatomy departments. The Visible Human Project (VHP) in America was inspired by this to generate a comprehensive set of transverse CT, MR, and cryo-sectional pictures of the whole human body.

A male and female cadaver was cryo-sectioned full length at one or one-third milli-meter intervals. Representative CT, MR, and cryosection pictures were also recorded. The VHP project was begun in 1986 by the US National Library of Medicine. The male subject's data set was completed in 1994, while one of the female subjects' data set was completed in 1995.

This extraordinary collection, which is on exhibit at the National Museum of Health and Medicine outside Washington, DC, is a watershed moment in medical imaging and anatomy instruction.VHP was duplicated in 2002 as the Korean Visible Human Project and the Chinese Visible Human Project to capture even better resolution photos (2002-03).

These cross-sectional pictures are turned into 3D accurate images of anatomic structures using volumetric reconstruction. These collections of 3D anatomical photos have resulted in a priceless library for medical study and education. Virtual surgery and virtual medical procedures like endoscopy, lumbar puncture, and cardiopulmonary resuscitation were developed as a result of the introduction of 3D renderings via virtual dissection software.

For medical education and clinical practice, cross-sectional photographs and digital virtual human anatomy models were made available. Anatomage, Bio-digital, Netter3DAnatomy, Visible Body, and Primal Pictures are just a handful of the medically accurate 3D anatomy tools and platforms that have transformed anatomical study.

These research systems hold more promise than cadaver-based investigations because they avoid issues like structural recognition, expense, repaired delicate structures, ethics, and hygiene that come with dissection. The virtual human is an interactive computer model that incorporates a human's morphological, biophysical, physiological, and biomechanical properties.

3D printing of human anatomy models is one of the most recent ambitions in anatomy. In 1986, Charles Hull created 3D printing technology to allow quick prototyping of plastic items. To make an object, 3D printing, also known as additive printing, layers a variety of printing materials based on a computerized 3D object model. 3D printing technology allows for the creation of intricate 3D replicas using a variety of polymers, metals, ceramics, and other chemicals.

Applications for 3D printing have grown in popularity, ranging from simple 3D model-based anatomy instruction to surgical practice and complex regenerative biology research. 3D printing is changing new medical applications and addressing scientists' and engineers' current needs for biomaterial scaffold creation.

Better 3D depiction of complicated anatomical systems has resulted from the collecting and analysis of realistic 3D anatomical information, as well as sophisticated software-based 3D rendering. Blender, Sketchup, Solidworks, Autocad, Maya, Meshmixer, and Rhinocerous are examples of advanced applications that create printable designs, which is an important first stage in the 3D printing process.

Bioprinting is a method that involves 3D printing biocompatible materials, cells, and supporting components into complex 3D functioning living tissues. 3D bioprinting, however still in its infancy, has high promises for addressing the need for tissue in organ transplantation and regenerative medicine.

Several hospitals are already using 3D bioprinting to create multilayered skin, bone, vascular grafts, tracheal splints, heart tissue, and cartilaginous structures, to mention a few. Human illness models are developed using 3D bio-printed meso- and micro tissue/organoid models for high throughput screening in research, drug discovery, and toxicity.

In 3D bioprinting, several considerations must be taken, such as material selection, cell type sensitivities, growth and differentiation factors, and technical obstacles related to tissue niche mimicking. To address bioprinting issues, knowledge from a variety of domains is required, including engineering, biomaterials science, cell biology, physics, and medicine.

Virtual reality-based visualization software now allows doctors to digitally exhibit distinct anatomical images in order to better communicate patient-specific disease symptoms. 3D printed anatomic models have been used as high-quality teaching aids for medical and surgical training, as well as clinical tools for surgical pre-visualization and implant creation in reconstructive surgery, in recent years. Virtual reality, holographic representation, and Google Glass can give 3D realistic visualization and a better understanding of anatomy and physiology new dimensions. The increasing complexity of 3D printing and virtual reality will pave the path for future regenerative medicines, improved medical education, and safe medical procedures.

1.5 ANATOMY AND ITS IMPACT ON MEDICINE

Anatomy is the study of the structure of the human body. Gross or topographic anatomy is what is visible to the naked eye, and it is the subject of this editorial. Two approaches will be used to consider the impact of anatomy on medicine. The first approach is from the standpoint of research: anatomy as an area of knowledge that has been gradually modified as a result of research discoveries.

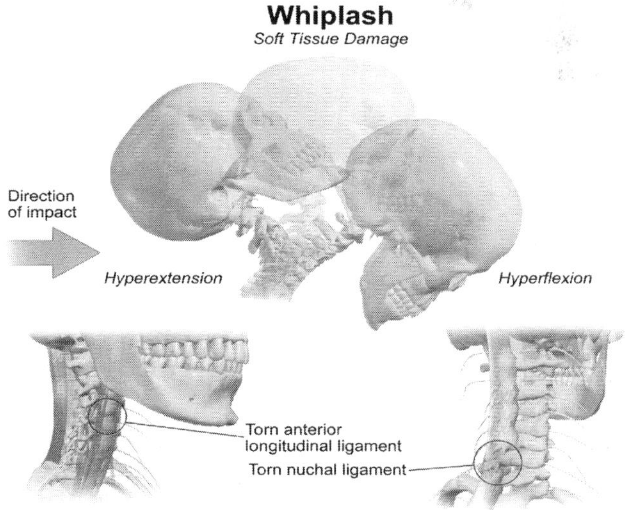

Figure 1.11. Image showing whiplash.

Source: Image by Wikimedia commons

This poses a series of related questions. Is anatomy becoming extinct as a topic of study? Is anatomy, on the other hand, on the verge of a renaissance? If so, what are its most recent and likely future developments, particularly those that coincide with medical advances? These are examined in the context of past, current, and future anatomical studies.

The second approach is from a teaching standpoint: anatomy as a topic. This, in turn, raises a series of questions. Is anatomy instruction in danger of becoming extinct? Is anatomy education, on the other hand, constantly developing and renewing itself?

If true, what are the present and likely future educational innovations, as well as their consequences for both undergraduate and postgraduate specialty trainees in medicine? Similarly, while addressing anatomy education in the past, present, and future, these are considered.

1.5.1. Anatomical Research (Past)

Much of what we know about medicine appears to be assumed. However, it's helpful to put this into context by citing specific examples of big situations and their remedies. Access, bleeding, infection, and discomfort were all major concerns in the early days of surgery. Anatomy, the ligature, antisepsis, and analgesia were among the advancements.

Despite the fact that anatomical knowledge is a prominent and vital component of medical study and practice today, it was crude in medieval times. The typical instructions used by medical practitioners were primitive squatting figures with practically unrecognizable organs. Leonardo da Vinci's anatomical studies allowed him to have a better understanding of how the human body functions.

They also provided insight into the structures underlying the surface patterns (particularly the musculature), which had been erroneously depicted by artists in the past. Interestingly, descriptive anatomy as a discipline did not emerge until Vesalius' De Humani Corporis Fabrica was published in 1543 (24 years after Leonardo's death). That work sparked a rush of interest in documenting human structural detail, which began with thorough cadaver dissection.

1.5.2. Anatomical Research (Present)

Isn't it true that every portion of us visible to the naked eye has already been discovered? Even if the human body does not appear to change, the ways in which we view, conceptualize, and intervene with it do. Recent advancements in surgery, as well as the rise of other anatomically related professions (especially medical imaging), have allowed alternate access routes, reduced invasiveness, averted tissue rejection, and permitted selective pain control.

Interventional radiology, endoscopic surgery, transplant surgery, and anesthetics are just a few of the domains that have emerged as a result of these advancements. Special imaging techniques such as computed tomography scans, magnetic resonance imaging, and ultrasound are enabling new ways to observe the body. Reconstructive and microsurgical procedures, anastomoses, implants, and robotics are among the other interventional improvements to emerge.

All of the foregoing advancements have been linked to parallel research aimed at identifying and describing the look and function of the important anatomical structures that have come into focus as a result of the various perspectives. Tracing the direction of relaxed skin tension lines, as well as identifying zones and circulatory segments of solid viscera, are only a few instances (particularly prostate, liver, kidney, and lung).

On endoscopy, it was required to distinguish between the appearance of normal hollow organs, body cavities, and joint cavities. Similarly, establishing the appearance of normal tissues on ultrasound imaging was essential when using ultrasound guidance for peripheral nerve and plexus blocks. What will happen when every primary care physician has access to a diagnostic ultrasound machine?

Although technology allows us to see increasingly smaller details within the human body, we must still interpret what we see. Certain previously unseen structures will be difficult to spot, while others may be misinterpreted for pathology. As a result, understanding anatomy is more critical than ever to ensure that we don't see too little or too much.

Anatomy is more than just names, labels, and descriptions. Angiosomes for vascular supply areas is one of the new concepts being developed. The angiosome notion is, in my opinion, the single most significant anatomical advancement in the last 100 years. This has revolutionized plastic surgery by allowing free grafts to be performed with a high chance of success.

The enteric nervous system is another major conceptual breakthrough that is revolutionizing the diagnosis and treatment of gastrointestinal motility disorders.

1.5.3. Anatomical Research (Future)

What's in store for the future? More highly targeted procedures will become available, especially as surgical, imaging, and anesthetic techniques continue to progress. We must be mindful of anatomical variances that may have an impact on new findings at the same time. As a result, looking forward necessitates looking back.

In other words, we require a windscreen-mounted rear-view mirror. This necessitates examining earlier documentation of such variants (anomalies), which have been recorded in great detail, but not in the context of hitherto inconceivable surgical operations. Furthermore, it is becoming increasingly vital for researchers to have access to cadavers in order to test surgical techniques (preferably) before performing them on actual patients.

The presence of an unexpected anatomical variance can cause complications following a new surgery. Rather than abandoning an otherwise sound surgical breakthrough, the dissecting room might be used to fine-tune the procedure to accommodate for this likelihood. Who'd have guessed that cardiac surgeons would be thinking about changing the artery supply to the hand?

When harvesting radial or ulnar arteries as potential ducts for coronary artery transplants, this is critical. Future studies will presumably reveal relationships between specific anatomical variants, which will aid in predicting which ones are most likely to be present. Anatomical research must also continue to challenge previously held beliefs and dogmas.

Similarly, clinicians must be cognizant of the consequences of their actions. Surface anatomy is a sobering example of something that needs to be updated in light of data from living subjects using contemporary imaging tools. Inconsistencies exist within and within anatomical reference materials, which are generally focused on cadaver research rather than *in vivo* evidence.

Some widely used surface markers and anatomical planes were determined to be accurate, whereas others were not. Gender, age, posture, breathing, build, and ethnicity have all been shown to have an impact. Older notions must be redefined in tandem with the introduction of new ones. Prior maps have been demonstrated to be erroneous due to being developed from defective or insufficient investigations, in contrast to a suggested new evidence-based dermatome map.

Different sensory modalities (touch or pain) with varying degrees of overlap can be represented using dermatome maps. Dermatomes are depicted as flat in both the cases; however, they should be renamed neurosomes to reflect their three-dimensional nature (like angiosomes).

1.6 CONCLUSION

Thus, it can be concluded that "Anatomy" is the most basic term and area of study in the life sciences. Anatomy usually refers to the internal and external structures of the body and their physical relationships. Herophilus first laid the factual groundwork for gross anatomy, which is the study of structures large enough to see even without a microscope. Ideas of Galen were the authority for anatomy in Europe until Andreas Vesalius's methods came on a firm foundation of observed fact. The microscope later allowed the discovery of tiny structures such as capillaries and cells, the subject of microscopic anatomy. Crucial advances in this field, including the microtome, which slices specimens into extremely thin sections, and staining, led to the new fields of cytology and histology.

REFERENCES

1. Eizenberg, N., 2015. Anatomy and its impact on medicine: Will it continue? Australasian Medical Journal, [online] pp.373-377. Available at: <https://www.ncbi.nlm.nih.gov/pmc/articles/PMC4701898/> [Accessed 18 June 2022].
2. Encyclopedia Britannica. n.d. bioclimatology | science. [online] Available at: <https://www.britannica.com/science/bioclimatology> [Accessed 18 June 2022].
3. Encyclopedia.com. n.d. Plant Anatomy | Encyclopedia.com. [online] Available at: <https://www.encyclopedia.com/social-sciences/applied-and-social-sciences-magazines/plant-anatomy#:~:text=As%20a%20part%20of%20botany,flowers%2C%20fruits%2C%20and%20seeds> [Accessed 18 June 2022].
4. Karl Gegenbaur, 2022. Karl Gegenbaur | German anatomist. [online] Encyclopedia Britannica. Available at: <https://www.britannica.com/biography/Karl-Gegenbaur> [Accessed 18 June 2022].
5. Körperwelten. n.d. The History of Anatomy - from the beginnings to the 20th century. [online] Available at: <https://bodyworlds.com/about/history-of-anatomy/#:~:text=3rd%20century%20B.C.,animals%2C%20primarily%20pigs%20and%20monkeys> [Accessed 18 June 2022].
6. Paul, M., 2019. Anatomy: Past, Present and Future. Journal of Human Anatomy, [online] 3(1), p.4. Available at: <https://www.researchgate.net/publication/333313549_Anatomy_Past_Present_and_Future> [Accessed 18 June 2022].

CHAPTER 2

HUMAN ANATOMY - REVIEWS AND MEDICAL ADVANCES

CONTENTS

For others, anatomy is defined by formalin cadavers in antiseptic dissecting rooms and massive quantities of precise anatomy. It has long been viewed as the mainstay of medical education, even as a handmaid of surgery, inside academia. Others see it as having a far more humane aspect, as seen by ceremonies at the beginning and conclusion of dissection to unite the dead corpse with the once living people and their relatives.

It is conveyed to the general public via tremendously popular public exhibits of plastinated cadavers and body parts. However, these definitions are only elements of what comprises modern anatomy, with its strong research spirit and vast breadth ranging from biological anthropology and clinical anatomy to molecular genetics and biology.

The underlying thread that connects all these techniques within current anatomical research is structure or organization, which is the one unifying topic throughout this large swath of biological activity. From 1858 until the present, these trends may be tracked through the several editions of Gray's Anatomy.

The humanistic aspect of anatomy, on the other hand, reminds us that anatomy is more than just a science. Its ethical issues involved are just as numerous just like its scientific credentials since it has evolved from a debatably ethical and practical action to one which presently seeks to express the highest standards of ethical conduct, even though it still has challenging dimensions in many societies, such as its continued reliance on the use of unclaimed bodies.

Anatomy has entered the unknown ground in these and other ways, with hitherto unknown moral considerations.

2.1. HUMAN ANATOMY: A REVIEW OF THE SCIENCE, ETHICS AND CULTURE OF A DISCIPLINE IN TRANSITION

2.1.1. Setting The Scene: *Gray's Anatomy*

Gray's Anatomy, the anatomists' standard work, reflects the different facets of anatomy. The 41st edition of the English translation, released in 2016, 156 years after the first, is both spectacular and nearly thorough in its scope. Its key sections vary from Cells, Tissues, and Systems to geographical subdivisions of the human body.

The reader is left with a visual impression of high-quality pictures that rely on a variety of modern methods, ranging from traditional histology to immunofluorescence and immunolabelling. This is the modernized method of conventional anatomists.

However, most of the detail is light years away from what was accessible even 50 years ago, let alone 150 years ago. The nickname, the 39th edition of 'The Anatomical Basis of Clinical Practice' was published in 2005, emphasizing that anatomy should be understood from a health sciences perspective since this is a primary motivator for understanding and appreciating the anatomical arrangement of the human body.

It is also worth noting that, beginning with the 1973 edition, there has been a readiness to acknowledge that there are gaps in our knowledge of anatomy, especially gross anatomy and that our understanding of the human body is far from complete.

In other words, continual research is essential if anatomy at all levels is to keep up with discoveries in related biomedical fields, a theme that following editions of Gray's Anatomy have endeavored to carry on.

Whereas a new edition of this classic literature may enthrall readers, its origins may be traced back to the 1850s in mid-Victorian London. It was a time of tremendous cultural turbulence in both the literary and scientific domains. Surgery, too, was making enormous progress as anesthetic became more widely available.

When two young medical professionals, Henry Gray and Henry Vandyke Carter, who was also an exceptional artist, began working on the project that would end in the publishing of Anatomy Descriptive and Surgical, it was a moment of excitement and ferment. Little is known about Gray himself, save that he was a promising physician who died of smallpox three years after his Anatomy was published, just at age 34.

Two points are important to mention. Gray was indeed a scholar, for starters. Gray had written studies on the histology, embryology, and comparative anatomy of the optic nerves and, subsequently, the spleen. The first book, The Structure and Functions of the Human Spleen, was based on this latter work. Research was an important foundation for the work he did on his magisterial message a short while later.

Second, Gray and Carter performed the dissections themselves at St. George's Hospital in London. The dead would also have been destitute people from workhouses, jails, and hospitals whose remains had not been claimed by families. They went unclaimed, yet given the prevalence of wrongdoing and the lack of official records, it's likely that several went unclaimed due to fraud.

Their origin or dilemma is never stated in Gray's Anatomy. Although this is much more usual than unusual in anatomical texts from the mid-nineteenth century and beyond, it indicates a disconnect between the gorgeous drawings of normal human anatomy and the bodies that furnished the raw resources for the representations.

"In Gray's, the legally sanctioned bodies of persons entirely alone in the metropolis were the raw material for dissections that served as the basis for images that were represented in print as wood gravings," historian Ruth Richardson writes.

They have entered the brains of generations of live people as mass-produced images—via the eyes, minds, and ideas of those who have stared at them." There is no memorial to individuals whose bodies supplied so much for generations of anatomists and anatomy trainees.

There seems to be no proof that these two men did anything improper. Nonetheless, they serve as a reminder that the milieu in which modern clinical anatomy was developed was far from twenty-first-century aspirations.

By the same token, they were a long way from the civilization that gave birth to anatomy as a modern business three centuries earlier, during the European Renaissance. As a result, anatomy, like every other study, is a living thing. Anatomy today should be different than it was 50 years ago because it will be changed again in 50 years. Anatomy takes on diverse shapes in different places.

2.1.2. Anatomy and the Culture of Dissection

Anatomy as science evolved throughout the Renaissance as it sought to carve out its own position among the growing number of academic subjects. If the word 'anatomize' and noun 'anatomisation' were used today, they would be in a scientific context.

This is since anatomising or dissecting a body reduces it to its component pieces in order to develop a new body of knowledge. The goal is to enhance and widen the knowledge of anatomy in general based on what is learned about individual bodies.

Even though the term "body" is commonly used in reference to the body of a human or an animal, it is an abstraction. We are constantly shifting between a specific body, someone else's body, and the body in general. Cadavers, body parts, tissues, and bones remain are always from specific people, and even though these people lived in the distant past, they can never be entirely dehumanized.

The culture of dissection arose from a dizzying combination of opposing cultural factors in sixteenth and seventeenth-century Europe. Anatomy was part of a popular culture captivated with the interior of the body but unable to dive into this mostly undiscovered subject prior to the contemporary, impartial, scientific approach to the human body.

There had been considerable interest in the body, which evolved a morbid obsession with dissection because this was the sole means to explore the interior of the body. Moreover, because the bodies were fresh off the gallows, there was even a tight relationship between both the anatomist and the executioner.

As a result, the criminal, the executioner, and the anatomist all played a role in what has been dubbed "the culture of dissection." In this horrific process, the three were intertwined, with no sense or trace of clinical separation.

As a result, anatomists became active players in the execution process, even appearing to be accomplices of the executioners. This was necessary to ensure that the supply of human corpses did not run out. This, however, was an issue since it gave a clear message to the general public that anatomists were far more like criminals than respectable members of society.

To overcome this, their actions were imbued with heavenly importance, and when they probed the normally inaccessible world of the body's innards, they were peering into what was effectively a sacred temple. In this way, anatomists' standing was elevated to that of divine acts, raising them above the level of criminals.

The concept of self-dissection, exemplified by Andreas Vesalius and others, was a similar phenomenon. As the title indicates, the dead corpse was shown as actively participating in the dissecting procedure.

This suggested that the body was not as dead as one would think, since anatomy was animating it and giving it a life of its own. As a result, the perception was created that knowledge of the (dead) body was true knowledge of the living, emphasizing the naturalness of dissection.

Despite these small changes, cadaver dissection remained difficult. It was still on the outskirts of live civilization, with dissected cadavers perceived as a frightening community of the dead. Vesalius and others claimed that by showing cadavers in this manner, the anatomist was not upsetting the corpse, but rather facilitating the natural process of decomposition.

All were attempts, in different ways, to bridge the gap between the dead and the living, actions that have recently resurfaced in public exhibitions of plastinated bodies (see Anatomy exposed to public gaze—plastination). This implies that civilizations' aversion to dissection and the use of the dead corpse persists, regardless of how far the circumstances around here have changed.

Figure 2.1. Public exhibitions of plastinated bodies.

Source: Image by Flickr

For anatomists, the outcome has been a conflict between the legitimate scientific urge to operate on high-quality material (fresh material received shortly after death) and the ethical requirement of obtaining informed consent from a donor prior to death.

While not all uses of unclaimed bodies fall into the outrageous categories mentioned above, and while the precise nature of informed consent is still being debated, the absence of any reference to the importance of informed consent has cast a pall over the ethical environment of anatomy as a discipline.

This will only be corrected if anatomy explicitly advocates and executes the ethical superiority of body inheritances over abandoned bodies.

2.1.3. The Humanistic Face of Anatomy

Up to this point, the debate has focused on how anatomy has achieved a footing in the scientific community, giving it a legitimate stake in inquiries into the structure of the human body. This underpins its potential contributions to research and education.

The avenues accessible to anatomists in research inquiries are wholly reliant on their scientific qualifications, and this has also been the case in teaching the principles of anatomy, at both the macroscopic and microscopic levels, until recently.

However, it has become clear that restricting anatomy to its scientific dimensions in educating health science students has limits, as these students will be joining professions where empathy for patients is essential.

As this idea has been more entrenched in educators' minds, there has been a significant shift toward viewing anatomy as a humanistic as well as a scientific profession. This is not a denial of anatomy's scientific foundation, but rather an attempt to situate it within a patient-centered health science perspective.

One aspect of this is the rise of commemorations and memorials associated with body donation for medical training.

Ceremonies to recognize and honor those families of people who have contributed their bodies to anatomical learning demonstrate that anatomists are valued members of their communities who rely on the generosity of others. This conveys the idea that anatomists are human beings working with the remains of fellow humans, and that anatomy as a study takes this relationship into account.

These rituals openly state that the corpses available for study were given by persons who gave their full informed permission. This, in turn, underscores the importance of bodily bequests in ethical considerations within the anatomy.

Commemoration, thanksgiving, ceremony, service, and memorial are all names used to describe the core of these rituals ('memorial ceremonies'; 'Convocation of Thanks', 'Thanksgiving Service'; 'cremation/burial ceremony'). Each portrays the idea of memory and paying respect to individuals who, in their death, devoted their bodies to a great cause, that of medical instruction and research, in their own unique way.

Contributors were remembered for how much they have given, as well as a ceremony is one public display of this priceless gift. Their generosity is recognized and celebrated, and their families are thanked for their assistance in making this gift possible.

Giving something more personal to oneself represents faith in the anatomy faculty and students, with the assumption that their bodies will be handled with care and dealt with in a manner reminiscent of the donor's memory.

2.1.4. Anatomy Exposed to Public Gaze: Plastination

Numerous authors have characterized and evaluated these massive plastinated exhibits and their place in the realm of anatomy. They are aimed at the general population rather than academics and students in medicine and other health sciences.

They are meant to bring anatomy out of the private dissection room and into the public eye. Gunther von Hagens, the creator of the Body Worlds empire, has referred to this as the 'democratisation' of anatomy, the liberation of anatomy from its privileged place within the halls of academia and into the wider world.

Everyone should have access to knowledge about their own body, and presenting the body in its dissected condition is the best way to do this. Only in this manner could individuals start to understand what organs are like and how they connect to one another, as well as the veins and nerves that supply them.

But how will this be accomplished, given that a copy of the dissecting room, with its scent of formalin-impregnated death and dead frozen cadavers on slabs, is unlikely to draw large viewers?

The major breakthrough accrued when von Hagens invented plastination, a revolutionary way of preserving human tissues in which tissue fluids are replaced with plastic.

This is a significant advancement for use in teaching human anatomy to health science students, as it is utilized to preserve previously removed body parts. It was also useful for scientific purposes, thanks to the usage of body slices. Of course, these applications are limited to academics.

The transition to public exhibitions coincided with the preservation of full bodies, known as plastinates, rather than body parts. However, most importantly, they are being shown vertically instead of horizontally. Maybe not all, but they may be shown in a number of poses, giving the sense of running, walking, leaping, participating in a variety of sports, and even engaging in sexual intercourse.

They look to be 'alive' when using these instruments, a vast cry from the dead passivity of the dissecting room. They may be disassembled, yet they give the sense of being involved in the vibrant activity of daily existence.

The result is usually spectacular and awe-inspiring, eliciting astonishment at the beauty and intricacy of the human body. For some, this favorable aspect is accompanied by utter rejection since they are a travesty: filthy, uncomfortable, degrading, and dehumanizing.

What's remarkable is that the original opposition to them came mostly from anatomists and religious leaders. While more in-depth investigations have removed some of this hostility, the passionate responses of certain people have shown sources of concern.

2.1.5. Anatomy and Its Ethical Dimensions

Recent breakthroughs in anatomical thinking have resulted in concentrated efforts to elevate the visibility of ethical thinking as a foundation for anatomical thinking and research.

While this has been performed by individual anatomists, anatomical associations representing anatomists from throughout the world have also taken on board. These associations are represented by the International Federation of Anatomists (IFAA), which has developed a set of ethical rules to regulate the donation of human bodies.

To provide donors complete trust in their decision to contribute, procedures of the highest ethical standards are essential. This, in turn, necessitates the public's faith. The rules are as follows: The basic concept is that bodies have been left to be used for teaching and study.

In summarizing the findings of her magisterial study of anatomy during the Third Reich, Hildebrandt referred to studies on the "future dead," as one moral principle after another was abandoned and the career was transformed into "a representative of evil through the convergence from their own reductive idea of human life, National Socialist exclusionary medical ethics, and the new "opportunities" provided by the regime."

This path included what she calls the "destruction of memory" and the total elimination of all professional ethics. The take-home lesson for Hildebrandt is that "the benefit to the person must remain at the core of medical ethics, not the possible benefit to society as a whole." In that regard, the doctor must take a political stand.

2.1.6. Future Dimensions of Anatomy

The essential achievements of Vesalius and Gray, among several others, must be honored on a regular basis for their seminal contribution to what anatomy is today. Nonetheless, they were children of their periods, working in surroundings very distinct from each other and from those faced in the twenty-first century. We can't comprehend them or their contributions until we grasp their surroundings.

Similarly, humans cannot be understood apart from our settings, and we also have the potential to modify them in at least some ways.

Anatomy somehow doesn't remain static, and neither do the expectations of one civilization from those of comparable and quite diverse nations. Moreover, historical lessons may be significantly more applicable to present situations than could have been predicted. For example, it must now deal with both the pressures and opportunities presented by cyberspace.

One of the primary worries is how the availability of anatomical dissections on media such as YouTube may standardize popular ideas of anatomy in ways that the anatomical profession has little control over. It must be seen if it will have an impact on the faith that really is essential towards the connection between institutions, donors, families, and communities, as well as the existence of good donor programs.

Technology is having a significant impact on anatomy, as it is on every other health science area. Among them are ethical implications, and if its practitioners do not understand this, the effects might be detrimental to human welfare.

2.2. INNOVATIVE TECHNOLOGIES FOR MEDICAL EDUCATION

Medicine is the study and practice of illness diagnosis, treatment, and prevention, and most of the medical information concerns our human body. Medical information is used in a variety of contexts, including education, training, diagnosis, surgery, and so on.

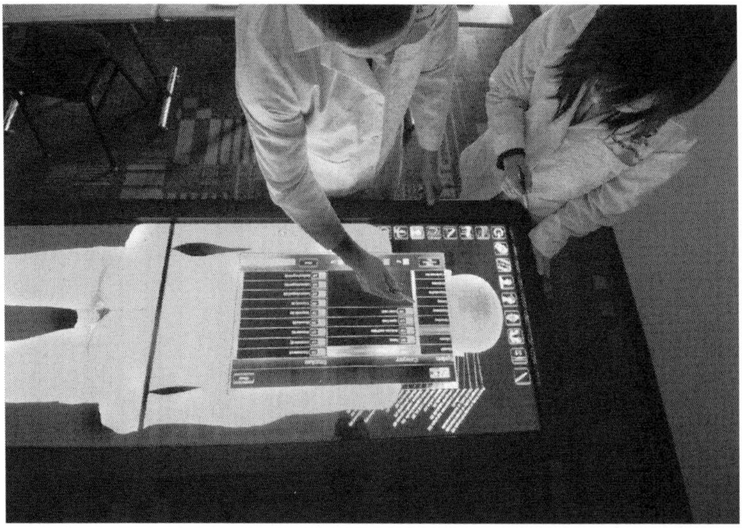

Figure 2.2. The Harker School Anatomy Table.

Source: Image by Wikimedia Commons

As computing technologies advance, more information from the physical world gets digitized, and researchers focus on ways to process and display it to the user, allowing the user to perceive and interact with the information organically and efficiently. The fundamental human behaviors of perception and interaction with various media and objects remain user-specific.

Anatomical education is an essential part of any curriculum and begins quite early in school in order to create a good understanding of the body and raise health awareness among the general public.

With the arrival of a slew of exciting technical breakthroughs, there should be no reason not to incorporate these into the development of new educational paradigms for s. As such, this chapter provides an overview of the intriguing fresh research being conducted in this field.

2.3. STATE-OF-ART

The expertise of a physician is accumulated during many years of medical practice. The information will be imparted to medical students through medical education and to the general public through public education in a variety of contexts. In the surgical environment, for example, medical imaging data is gathered for diagnosis and navigation.

Good communication between the patient and the surgeon is critical for ensuring that the patient follows the surgical procedure. The same communication is essential for patients who are doing rehabilitation activities following surgery. In each of these scenarios, critical actors perform various roles in order to perceive and engage with medical information.

In this section, we explore the current condition of medical education, public education, and communication to prepare for surgery and rehabilitation, as well as the problems involved.

2.3.1. Medical Education

Despite the fact that medical education has existed for millennia, it begins with Abraham Flexner's report in 1910. In 1909, Flexner toured 155 medical schools in North America and recognized four fundamental difficulties in medical education:

1) A lack of uniformity
2) A lack of integration
3) A lack of inquiry and
4) A lack of identity formation

The study had a major influence and helped define current medical education, which combines patient care, teaching, and investigation. Academic institutions, on the other hand, do not devote enough time to teaching due to great pressure to publish as well as economic considerations.

Another difficulty is that research is focused on relatively tiny subtopics that have little to do with medical education. On the other side, limited research is conducted in the field of education. For example, in gross anatomy, most issues are well-known and there is minimal new study.

Students in today's medical schools are expected to comprehend both the functional and geographical context of human anatomy.

Traditional methodologies:

Learning in traditional medical education is divided into three categories: corpse, model, and book. Despite major advances in technology in recent decades, school instruction continues to rely on the same ways to impart anatomical information. The material is often collected in printed publications, such as anatomy atlases, and shown in the form of charts and diagrams.

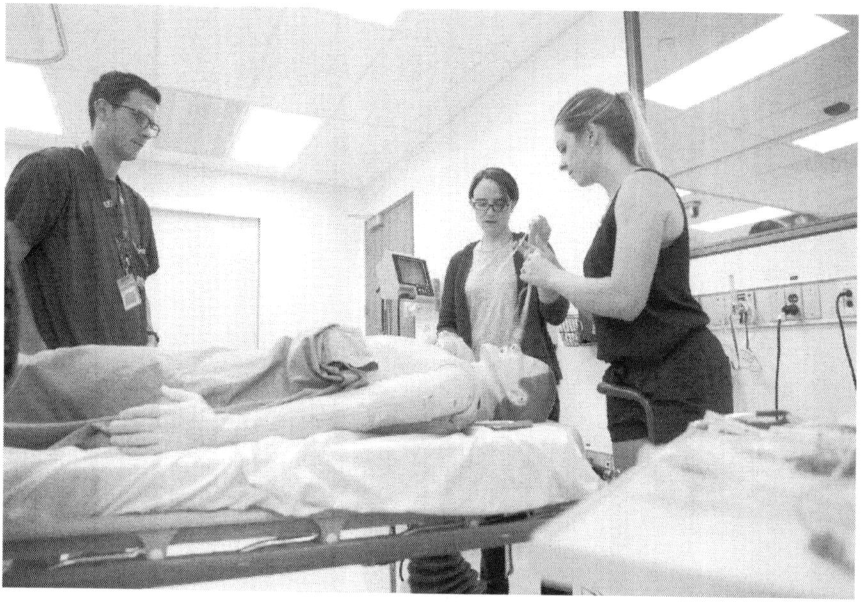

Figure 2.3. The traditional method of learning medical education.

Source: Image by Foundations of Educational Technology – Pu Pub

These diagrams give a straightforward and well-known approach to illustrating organ structure and appearance, with the benefit that the user is familiar with similar types of display. This strategy, however, has significant drawbacks.

First and foremost, the author has chosen to provide only a few diverse cross-sections. Because occlusions restrict the options for seeing these spatial interactions, this may not be adequate in some circumstances to completely represent how an organ is placed relative to its surroundings.

Another difficulty is that organs are frequently portrayed conceptually, missing out features or distorting tissue colors, providing only a rough picture of how the organ truly looks in reality.

It is difficult, for example, to grasp the spatial and physical aspects of anatomy by looking at two-dimensional pictures, diagrams, or photographs. Many physical models also lack the amount of detail required to completely comprehend the particular anatomy.

***Computer-based learning**:*

It is created by specialists, and learners may utilize it if there is no expertise accessible inside the hospital. Computer-based learning may be highly effective for anatomy instruction, especially when 3D visualization is included. There are several virtual model databases available, and E-learning is widely employed nowadays.

These are wonderful resources that are more engaging and fascinating than textbooks. Furthermore, customization in E-learning systems has been the focus of many recent studies, allowing teachers to pick and combine factors in order to develop alternative tailored tactics based on the particular of the courses.

In recent years, digital technologies that aim to improve on existing strategies have emerged. Several of them provide interactive anatomy models that may be accessed either online or as a standalone program. There are also organizations that specialize in providing teaching packages for use in schools, allowing alternative teaching techniques to be used at various school levels with the assistance of films and interactive tools.

2.3.2. Public and Patient Education

Medical school teaching methodologies are unsuitable for public education since a high degree of anatomical comprehension is not required. Because real-life demonstrations are generally restricted to the skin's surface, most individuals study anatomy through charts or plastic models of the inner organs.

Still, there is a certain curiosity with peering within one's own body, which is normally only possible to a limited extent through the use of X-ray imaging or comparable imaging modalities. These approaches are unsuitable for public teaching, owing to the high cost of the gadgets when used without a medical justification.

Patient education is a means through which health practitioners provide patients and their careers with the knowledge that will change their health behaviors or enhance their health condition. It can help patients better grasp their medical condition, diagnosis, disease, or handicap. Patient education is an essential and much-overlooked job of a healthcare worker.

A doctor's role is to inform and motivate patients so that they understand the illnesses, the treatment choices available, and the implications of not receiving treatment or failing to comply. This makes it easy for patients to comprehend it.

Ultimately, individuals will be able to take responsibility for their own health. Good knowledge and communication boost the patient's ability to participate in decision-making, resulting in increased levels of patient satisfaction, physician loyalty, and favorable treatment results.

Patient-health professional communication in a clinical context has traditionally consisted of a face-to-face narrative engagement, typically accompanied by static visuals or real-time drawing. Due to the complexity of medical knowledge, it is unavoidable that patients' comprehension of explanations is frequently weak.

Evidence reveals that patients commonly do not grasp what is spoken to them during a medical visit owing to cultural and educational differences between practitioners and patients. This discourages the patient and makes them unduly reliant on the advice of their doctor. Improving and improving the patient's information process and comprehension has been a focus of medical study.

Ong et al. summarized the important subjects of information exchange, interpersonal connection development, and shared medical decision-making in a literature review of doctor-patient communication.

Wilcox et al. presented a concept for patient-centric information displays in order to provide helpful information to a patient during an emergency department visit, as patients are frequently misinformed. Previous research has also shown that technology may improve communication.

Hsu et al. investigated the influence of in-room access to electronic health data on doctor-patient interactions during outpatient visits in longitudinal research. Recent evaluations demonstrate a growing body of data supporting the use of multimedia technologies to increase patient satisfaction and retention of information.

2.3.3. Rehabilitation Exercises

The procedure of regaining full function after an injury is known as rehabilitation. This entails regaining strength, flexibility, endurance, and power via a variety of workouts and drills. Rehabilitation is just as vital as care after an accident, yet it is often disregarded.

Physiatrists are typically involved in the exercise procedure, and ideal physical activities are carried out in an effort to achieve certain health goals. Its goal is to restore normal musculoskeletal function or reduce pain caused by injuries.

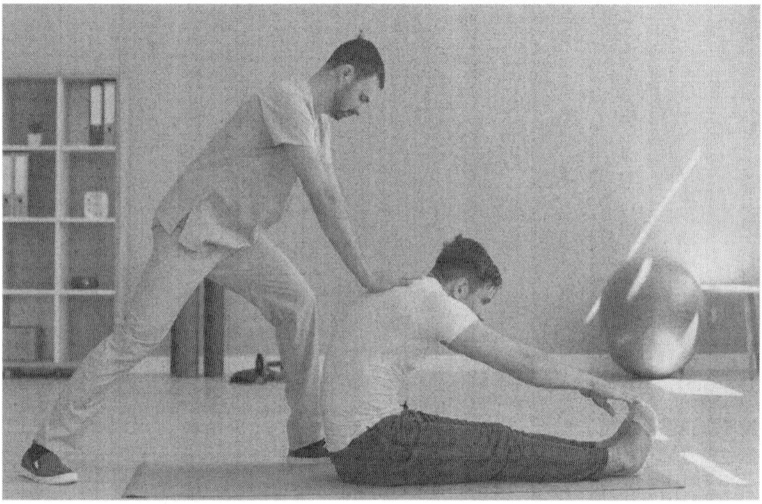

Figure 2.4. A patient doing rehabilitation exercises.

Source: Image by Midland Healthcare

Communication between a medical assistant and a patient is critical throughout the exercises, for example, ensuring that the patient is aware of his or her present level of incapacity and follows recommendations during the rehabilitation program.

Rehabilitation practices are often conducted at a rehabilitation facility, and the physiotherapist analyses and assesses the damaged motions and motor functions. Patients are advised to repeat the same workout motions multiple times in order to acquire kinetic gains as rapidly as feasible. Patients' learning through movement repetition is critical to their therapy's effectiveness.

2.4. THE MAGIC MIRROR

Delivering an acceptable learning experience to diverse learners is a difficult issue since traditional methodologies often cannot alter information to meet the demands of the individual learner. Personalization for supporting a multimodal learning environment is a rising topic of interest, such as the creation of user modeling and tailored procedures that put the student at the center of learning growth.

Augmented reality systems may display a virtual version of the subject material and provide a direct link between the knowledge the user wishes to learn and their actual body, as well as activities. As a result, it might aid in the comprehension and retention of difficult material, supplementing or even replacing traditional learning methods.

Earlier mixed-reality systems for anatomy visualization relied on expensive technology such as head-mounted displays (HMDs) or markers. A Magic Mirror technique for anatomy instruction has recently been created, which employs a camera for tracking and a TV display for visualization of an augmented reality perspective.

The described solution is both affordable and simple to operate. It displays medical data augmented onto the user's body and displays extra 2D and 3D information based on the user's need. The Magic Mirror grants the users 'the superman power' to see inside their own body.

It allows medical information to be interpreted intuitively by associating it with a genuine human body. In the Magic Mirror system, the natural gesture is chosen as the interaction mechanism, and interaction with the augmented reality image of the user's body offers a personalized perspective.

The Magic Mirror is a user interface approach that resembles a standard mirror and provides nonphysical visual input in addition to the optical effect. The user stands in front of a screen, and the image of the user is projected onto the screen through a camera, acting as a mirror. Previously, systems added things to the user; however, this system expands the notion of medical education and rehabilitation.

The Magic Mirror enhances the volumetric visualization of a CT dataset onto the user to produce this view. The user's stance is monitored to enable accurate augmentation of medical data. The Magic Mirror concept emerged as a system that produces a tailored impression of medical data for each user.

Furthermore, the system uses natural gestures from the user to build an interactive mixed-reality experience. Knowledge of human anatomy is essential for anybody working in the medical industry. This is also an essential aspect of general education and is applicable to several other professions, such as health care or sports.

Human anatomy is extremely complicated, requiring an understanding not only of a single organ but also of chemical processes, human mobility, and spatial interactions throughout the body.

As a result, teaching human anatomy is extremely difficult, and much effort is typically expended on it, for example, by allowing students to participate in dissection courses, generating pictures, using plastic models of anatomy, or employing 3D computer graphics.

The Magic Mirror system is originally used to display anatomical structures overlaid over the user's body in order to educate human anatomy naturally.

2.4.1. Hardware Setup

The Magic Mirror masters the ability on a few key abdominal organs, including the liver, lungs, pancreas, stomach, and bones. The system prototype gives the user a mirror-like impression by projecting a 'seeing glass' on the body and displaying the user's skeleton, generated using CT data and anatomical 3D models.

The system follows the motions of users by utilizing a depth camera and an algorithm to determine the user's stance from the depth image. This is accomplished with the Microsoft Kinect, which was initially designed to provide motion control of computer games. The Magic Mirror metaphor leads the user to imagine that he or she can see into their own body.

Simultaneously, medical data (CT, MRI, and a completely segmented collection of cross-sectional pictures of the human body) is enhanced in real-time. The present technology also displays static anatomy to the user and provides a simple user interface for selecting CT, MRI, or photographic slices.

An illustration of the hardware configuration may be seen here. The system's initial component is a display device. Large TV screens or video projection onto a flat surface have been employed in various configurations of the technology. The second component is a color camera that is situated adjacent to the display surface and looks at the user, allowing visual perception of the data.

The third element is a depth sensor that is situated beside the color camera and also has a comparable field of view and viewing direction as the color camera in order to capture information about the user's skeleton.

The current system employs the Microsoft Kinect V1, which is comprised of color and a depth camera housed in a mechanical enclosure. The depth sensor is an infrared camera that estimates depth values for each pixel by using structured light generated by an extra infrared projector.

Based on machine learning, the Kinect sensor can produce the user's skeleton and personal information. The color camera is used by the system to create a mirror-like effect for the user, and all nonphysical visual feedback is generated according to the user's skeleton and private information, and the relevant health data is rendered out to the human body via volume.

2.4.2. Medical Information

Apart from textual and 2D atlases, 3D volumes and models are key medical information representation approaches. In medicine, the most popular method is to employ 3D volumes, with one pixel or voxel for each point in 3D.

This is the type of information obtained through CT or MRI. The others are polygonal or surface models, which store simply the surface (for example, skin or the surface of an organ). Surface models are frequently used for educational models because they appear better.

While the Magic Mirror technology can make use of a CT or MRI scan, it is not necessary to obtain a CT or MRI scan of the user when it is not needed for medical purposes. As a result, we supplement the visible Korean Human (VKH) dataset's medical volume onto the user. This dataset is made up of a CT scan, an MRI volume, and a photographic volume that was created by stacking cryosections.

The majority of CT and MRI medical pictures are preserved in the DICOM standard, which is utilized in all institutions. Unfortunately, most research software does not support DICOM. The medical data for the Magic Mirror system is stored in an .MHD file. One disadvantage of volume data is that it cannot be warped in real-time.

As a result, if the user bends, this is not represented in the depiction of the medical data, resulting in incorrectly portrayed limb motions. Potential remedies to this problem have been published in Section 3.4, however, for present technology, which concentrates on the abdomen, this is a minor concern. It is more difficult to see structures other than bones on a CT scan.

In the initial effort, the CT volume segmentation was utilized to view distinct organs in the abdominal region. The visualization quality was poor because the segmentation lacks subpixel precision and transfer functions on CT intensities cannot produce realistic colors and textures of organs. As a result, rather than using volumetric data, extra polygonal models were merged.

The Anatomium2 collection contains polygonal representations of numerous human organs. The dataset yielded a scene graph containing many organs. Using Coin3D, this scene graph is superimposed on the user.

2.5. HUMAN BRAIN ANATOMY: PROSPECTIVE, MICROGRAVITY, HEMISPHERIC BRAIN SPECIALIZATION AND DEATH OF A PERSON

The typical brain weight in cerebrospinal fluid (CSF) is 50 g and 1400 g without CSF (the actual brain weight). The drop in brain weight is thought to be the result of CSF buoyancy or a microgravity environment induced by CSF.

The force of gravity can be defined in three ways in principle:

a) By use of acceleration or aerodynamic force,
b) Via buoyant force, which follows Archimedes' theory from 212 BC, which states that "any object totally or partially immersed in a fluid is buoyed up by a force equal to the weight of the fluid displaced by the object"; this is a weightlessness idea.
c) by a thing with no (or negative) mass or time (? soul). CSF buoyancy causes a decrease in real brain weight, resulting in a state of microgravity or weightlessness.

The pregnant uterus, which can have comparable effects, is an extension of this theory (buoyancy resulting in microgravity). The ratio of fetal growth to amniotic fluid volume is larger in early gestation than in late gestation. The fetus is in a flexed or antigravity (microgravity) posture throughout this time.

As a result, it might be hypothesized that an antigravity or microgravity environment is required for appropriate CNS development. During early gestation, this stage of development results in a flexed posture of the embryo (microgravity body position: just like the astronaut in space, curved or a horizontal position, whilst the gravity position assumes a vertical position).

Later in pregnancy, the fetus's microgravity posture shifts to a vertical gravity position, which is required for muscle and bone growth as well as birthing preparation (with gravity force: 1g or 9.81 m/s2).

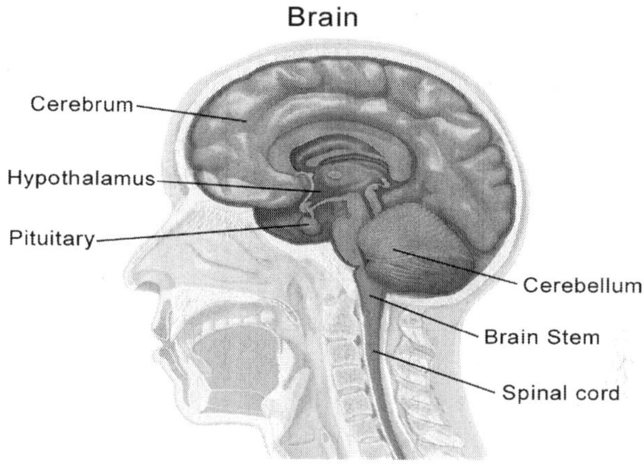

Figure 2.5. Human Brain Anatomy.

Source: Image by Wikimedia Commons

In terms of hemispheric specialization, the cerebrum is made up of two hemispheres joined by commissures, the biggest of which is the corpus callosum. In a healthily working brain, integration of information from both sides appears to be critical.

The use of magnetoencephalography (MEG) and electroencephalography (EEG) to analyze brainwaves and map the functional architecture of both hemispheres is demonstrated in this chapter. Other research has often used positron emission tomography (PET) or functional MRI (fMRI) to map and examine the functional and anatomical elements of language, sensorimotor, and auditory visual processes.

Researchers employed brainwave detecting technologies (MEG and EEG) to see the cortical brainwaves for the activities and examine their hemispheric activity and specialization.

Researchers also conducted a study of the literature on the anatomical structures involved in the rapid and efficient flow of information between the two cerebral hemispheres, such as the corpus callosum and other commissures, as well as a brief overview of callosal surgery.

Following reviewing the full brain as an organ in a microgravity environment, as well as cortical brain architecture and function (the surface portion of the brain), we finally examine the important topic of death, which has a tight anatomical relationship with structures in the 'deep and core part of the brain'.

Many ancient thinkers, including Plato and Leonardo Da Vinci, referred to this deep anatomical location as "the seat of the human soul" since it appears to play a critical role in either cardiac or brain death.

The above deep perivascular area encompasses anatomical structures of the brainstem, reticular system, hypothalamus, thalamus, basal forebrain or septal area, amygdala, hippocampus, and pineal and pituitary glands, and is also known as the 'greater limbic system,' a term coined by Nieuwenhuys et al. in 1988.

2.6. MICROGRAVITY INSIDE THE CENTRAL NERVOUS SYSTEM

a) The Archimedes buoyancy effect of CSF is related to the idea of microgravity within the CNS. Despite a small amount of CSF, buoyancy is maintained by:
    - The Windkessel phenomenon (vascular pulsations) that causes brain pulsation and thus well distributed intraventricular and extra ventricular cerebrospinal fluid that sandwich the brain parenchyma
    - the anchoring effect provided by the nerve roots, filum terminale, denticulate ligament at the bottom and cranial nerves as well as blood vessels at the skull base, and, most notably.

b) The brain is 70 percent water and 30 percent dry matter, with fat accounting for 60 percent of the dry matter.

In this regard, the evidence for the central nervous system may be found in the microgravity environment and is offered by:

- Weightlessness of the brain,
- Microgravity or bending posture at the midbrain level for the brain (as a result, words such as ventral and dorsal, rostral and caudal for the brainstem and spinal cord and cerebrum are different: For example, the phrase ventral for the brainstem is anterior, whereas ventral for the cerebrum is inferior, and the term

rostral for the brainstem refers to the superior end, whilst rostral for the cerebrum refers to the anterior end).

- During pregnancy, the chorionic and subsequently amniotic fluids create a buoyant environment for the development of the central nervous system.
- Sinking skin flap syndrome with altered cerebral blood flow in a chronic craniectomy patient
- The brain appears to float readily after open brain surgery,
- Brain shift after CSF buoyancy is removed: this may imply that the brain is in 'neutral buoyancy,' in which 'CSF density' is roughly equal to 'brain density,' and
- Research demonstrating that simulated microgravity improves mesenchymal stem cell development into neurons. These reasons suggest that the CNS may exist in a microgravity environment (between 0 and 1g or 9.81 m/s2).

2.6.1. Anatomical Relationship: Reticular Formation Network Anatomy, Microgravity inside the Central Nervous System and Origin for the Brainwaves

The cranial nerve, sensory relay nuclei, and ascending and descending fiber networks surround the classical reticular formation in the brainstem.

It is attached to all areas of the brain, including the neocortex (six layers of cerebral hemispheric cortex), archicortex (three to four cortical layers of hippocampus and olfactory cortex), and paleocortex (four to five cortical layers of rostral insular), parahippocampus, olfactory bulb, olfactory tubercle, piriform cortex, periamygdalar.

It is high in neuromediators such as noradrenaline, serotonin, choline, histamine, gamma-aminobutyric acid (GABA), and hypocretin. It is often separated into two systems:

a) ARAS (ascending reticular activating system) and

b) ARIS (ascending reticular inhibitory system)

These two divisions play a significant role in regulating awareness, the integration of autonomic (visceromotor), behavioural, and somatomotor responses, the endocrine system, and the control of the sleep-wake cycle. The reticular formation's components are exclusively identified in the brainstem, with links to the thalamus, hypothalamus, and basal forebrain nuclei (septal nuclei, etc.).

Nieuwenhuys and colleagues present an alternate picture of the reticular system, emphasizing its importance in limbic, hypothalamus, and parahypothalamic regions. They dubbed this new circuit the "greater limbic system," and recognized the hypothalamus, which is located rostrally beyond the conventional reticular formation, as an important component of it.

The traditional reticular development creates dispersed mosaic-like structures with several functional nuclei inside the brainstem, which physically comprises the medulla oblongata (myelencephalon), pons (part of metencephalon), and midbrain (mesencephalon). It is the center of the neuroaxis, which is physically oriented vertically or gravity-wise.

In contrast to the brainstem, the diencephalon, which includes the thalamus, epithalamus, subthalamus, hypothalamus, basal forebrain region, amygdala, hippocampus, and other periventricular structures, is horizontally positioned in an antigravity or microgravity posture.

The 'T' like shape of CNS cores and para cores is formed by a combination of these two postures. This is primarily due to the existence of mesencephalic or primary cephalic flexure during early brain development. If this flexure is not present, the brainstem and reticular formation will take a single vertical structure, with the hypothalamus thalamus constituting the rostral end.

This early embryological bending occurs as a result of the buoyant environment created by the chorionic and amniotic fluid during gestation and sustained throughout life by the CSF.

Intriguingly, a 1949 study by Moruzzi and Magoun revealed that the deep reticular system generates brainwaves and influences cortical brainwave rhythms via two pathways:

a) dorsal via the thalamus (thalamocortical network) and

b) ventral via the hypothalamus, basal forebrain region, amygdala, and hippocampus (extrathalamic network).

This extrathalamic network might explain why, in refractory epilepsy, peripheral vagus nerve stimulation can lower seizure rates (vagus nerve extrathalamic pathway hippocampus cortex). These two circuits go deep into the brain and form critical circuits (the CNS's core and para core) that deal with at least two major components of neurocognition:

a) Consciousness and
b) Memory.

2.6.2. Functional Relationship: Consciousness, Memory,and Origin for the Brainwaves

Consciousness and memory are regarded as two critical components of human cognition. Humans have a unique mental mechanism for obtaining information and understanding through cognition, experience, and senses.

This cognitive skill also enables some believers to appreciate creations and God (creator). If one's awareness or memory level has changed, it may be difficult to pray to God. As a result, one may argue that the CNS cores and para cores that give birth to awareness and memory are basically a seat of the human soul. Debates on the location of the human soul have raged since ancient times.

Plato (424–348 BC) and Galen (approximately 200/216 BC) identified the brain (encephalocentric theory) as an essential organ for the soul, but Aristotle (roughly 384–383 BC) identified the heart as an important organ for the soul. Who might have learned from Plato disagreed with Plato and favored the heart as the seat of the human soul?

Later, in the fourteenth century in Italy, Leonardo da Vinci (1452–1519) discovered the soul inside the brain, specifically in the middle ventricle close to the anterior portion of the third ventricle near the hypothalamus after drawing the intersecting infinity lines (golden ratio) of the human cranium.

The area identified by Leonardo da Vinci is actually part of the greater limbic system [it's worth noting that most structures in this deep anatomical area have infinite shapes, such as Solomon's knot (mosaic like reticular system), Pascal's spiral, Archimedean and Durer spirals (hippocampus, caudate nucleus), Lemniscate (thalamus) and pyramid (insular)].

As a result, it appears that the larger limbic system is an appealing concept for the home of the human soul for various reasons:

1. It is a location for 'brainwave origin.'
2. It regulates 'consciousness and memory' (two major components of human cognition that are strongly tied to remembering God); modification or loss of awareness (or memory) occurs when this deep region is harmed, making it impossible to recall or appreciate God.

3. A person's death would include this anatomical area—refer to the final part of this chapter for more information.
4. It may be regarded as the center of 'all brain networks' (at least one node arising from this deep brain area may be present in each brain network, and this node may look bigger than the others).
5. The 'infinity' lines of the skull intersecting here, as well as the majority of anatomical structures in this deep location, are most likely 'infinite' in their forms.

2.7. A STUDY ON HEMISPHERIC HUMAN BRAIN SPECIALIZATION

Pyramidal postsynaptic potentials, which have synchronous oscillations with the following, are the primary source of cortical brainwaves:

- Thalamocortical networks, which are modified by reticulothalamo cortical circuits.
- Extrathalamic cortical circuits, which primarily include the reticular system, hypothalamus, hippocampus, amygdala, basal forebrain, and septal nuclei and
- Additional cortices, known as cortical networks.

Magoun observed in 1952 that the reticular circuit in the brainstem plays an important part in the generation of brainwave patterns. The thalamus, hypothalamus, basal forebrain, and septal nuclei, parahypothalamic nuclei, pineal and pituitary glands, the limbic system, as well as the insula, basal ganglia, and neocortex, are all connected by this traditional reticular system in the brainstem.

The large interconnected networks generate optimum brainwave oscillations in the cortex via the thalamic and extrathalamic circuits, which may be examined using MEG and EEG.

In general, mapping the exact regions responsible for brain cognitive, sensorimotor, and auditory visual activities is difficult. Many people believe that these brain processes evolved deep inside the brain's center, involving physical locations that have extensive networks with the cortices.

The thalamus, hypothalamus, amygdala, hippocampus, basal forebrain and septal nuclei, reticular system, and pituitary pineal system comprise the central nervous system's core and paracore. These deep areas should have ideally been covered while mapping the areas involved in the functions.

Nevertheless, our research primarily looked at surface brain mapping and cortical brainwave analysis since MEG testing allows for more accurate, superficial, and non-invasive procedures than deep brain mapping.

2.8. A REVIEW ON CORPUS CALLOSUM, CALLOSAL SURGERY AND COMMISSURES

The corpus callosum is a large, transverse bundle of myelinated nerve fibers that connects the right and left hemispheres of the brain. It is anatomically separated into five regions: rostrum, genu, body, isthmus, and splenium.

It has already been proposed that this linkage and anatomy segment seem to be modality specific; the anterior callosal fibers connecting the frontal lobes transfer motor information, whereas the posterior fibers connecting the parietal, temporal, and occipital lobes bilaterally integrate somatosensory (posterior midbody), auditory (isthmus), and visual (splenium) information.

The corpus callosum develops from anterior to posterior, with the genu developing first, followed by the body, isthmus (characterized by a little narrowing at the level where the fornix abuts the callosum), splenium, and rostrum. It develops from the telencephalic alar plate's top segment through four stages:

a) prosencephalic cleavage (28–35 days of gestation),

b) commissural plate development (36–73 days of gestation),

c) corpus callosum formation (74–115 days of gestation), and

d) corpus callosum expansion (after 115 days of gestation).

The prosencephalon divides into the telencephalon and diencephalon during the prosencepalic cleavage stage. Following that, the single telencephalon gives rise to the creation of two telencephalic vesicles and a floor between them known as the lamina terminalis.

The lamina terminalis thickens and becomes known as the lamina reuniens or commissural plate during the commissural plate development stage. The commissural plate continues to thicken, and by 73 days, the four structures listed below may be seen inside it:

- the future corpus callosum,
- the region of the future anterior commissure,
- the hippocampal commissure, and
- the septum cavum pellucidum.

The corpus callosum develops from 74 days onward from crossed cortical axons through the commissural plate region. Axons from distinct parts of the brain cross at 'different times,' resulting in diverse corpus callosum regions and functions.

In contrast to the creation of the corpus callosum, the maturation and myelination process begins from the posterior to the anterior. It appears postnatally in the splenium at around 4 months and in the genu at around 6 months. The corpus callosum matures at roughly 8 months of age and continues to develop over the first two decades of life through a steady rise in size.

These myelinated axons allow for the rapid propagation of brain impulses or waves, which are required for appropriate cognitive, sensorimotor, and auditory-visual functioning.

Indeed, abnormalities in the corpus callosum, particularly those associated with brain malformations and syndromic kinds of agenesis, are associated with deficits in neurocognition, neuro-behavioral, sensorimotor, and auditory-visual functioning.

These lines of thought suggest that the corpus callosum is an important structure for cortical and interhemispheric connection, suggesting a computational necessity for interhemispheric coordination in proper behavior, cognitive, sensorimotor, and auditory-visual activities.

Callosal surgery should indeed be conducted with caution and a thorough understanding of its anatomy and connections. The anterior interhemispheric transcallosal approach to the lateral and/or third ventricles should resect only the anterior part, the rostral body, and a portion of the genu, thus sparing the crossing motor fibers from the primary motor cortices in the anterior midbody and avoiding motor complications.

The posterior interhemispheric transcallosal route to the pineal area and the posterior section of the third ventricle is rarely employed. This method includes removing the splenium, which may result in tactile, auditory, visual, or emotional abnormalities.

Whenever relatives and friends watch certain patients, they may appear to be grossly intact and unaffected, but when particular neuropsychological tests are conducted following surgery, the abnormalities can be considerable. Verbal anosmia, double hemianopsia, poor processing of verbal information, apraxia, or agraphia of the left hand are some instances of these issues.

In contrast, resective callosotomy for intractable epilepsy caused by severe, medically intractable seizures with akinetic seizures or drop episodes will benefit from corpus callosum excision. Callosal division should be carried out in the manner outlined above.

Resection can be prolonged anteriorly till the rostrum, where the anterior commissure serves as an anterior limit and is best perceived when the two fornices converge. In situations in which the seizure result is poor, the resection should preferably be expanded posteriorly to include the anterior two-thirds of the corpus callosum.

This indicates that excision should include the motor fibers that run in the front and potentially a portion of the posterior midbody, which are at risk of causing lifelong motor impairments.

As a result, the posterior limit is more difficult to define and is typically led by the desired clinical effects (surgical goal), navigation system, body thinning (isthmus), and appearances of the fornices (the isthmus is the area where the fornix abuts the corpus callosum).

The anterior, hippocampal or forniceal, habenular, posterior or epithalamic, and supraoptic commissures are also known to cross the midline and connect the two cerebral hemispheres. The anterior commissure can be seen on either side, under the corpus striatum and in the temporal lobe material. It links the amygdala and temporal lobes and contains olfactory tract decussating fibers.

It is a component of the pain neospinothalamic tract. The hippocampus or forniceal commissure is the second biggest of the commissural connecting bundles that connect the two hippocampi by joining the two crura of the fornix. The habenular commissure, which is located in front of the pineal gland and joins the habenular nuclei on both sides of the diencephalon, is the next structure.

It is linked to the midbrain's pineal and interpeduncular nuclei. The posterior commissure is a circular band of white fibers that crosses the midline on the dorsal face of the upper end of the cerebral aqueduct.

It links the pretectal nuclei and is responsible for mediating the bilateral pupillary light response. Lastly, the supraoptic commissure, also known as the decussation, is a crossing within the optic pathway system that connects the two eyes to the two visual cortices.

Image fusion for deep brain stimulation surgery or radiosurgery frequently makes use of anatomical information of these commissures, particularly the anterior and posterior commissures. They are not currently used in reconstructive surgery; but, in the future, they may be acceptable white matter targets for brain stimulation to influence processes emerging from specific parts of both halves.

2.9. CONCEPT OF DEATH RELATED TO BRAINWAVES

Having known the genesis of brainwaves (deep brain region), cortical functioning, and quick hemispheric information transfer (superficial brain area), perhaps the notion of death will be easier to grasp. He will not die if someone cuts off his 'limb, hand, mouth, or face,' but if someone injures the core or deep section of the brain (the seat of the soul), or the cardiopulmonary system, he will die.

As a result, death appears to be linked to two major human organs: the brain and the heart. There are two sorts of fatalities based on this:

a) cardiac or circulatory death and
b) brain death.

It appears that in both types of fatalities, the anatomical area associated with the development of brainwaves or the broader limbic system is prominently engaged.

Brain death is characterized as the termination of all brain functioning. All points associated with brain death are essentially documenting dysfunction in the greater limbic system (or the seat of the soul), such as:

- conscious level,
- autonomic disturbances,
- absent brainstem reflexes,
- flattened cortical brainwaves (bihemispheric dysfunction), and
- disturbance in vital signs (noteworthy that these vital signs such as respiration, heart rate and blood pressure can be preserved by ventilatory support and medications in brain death).

A malfunction in the anatomical area that regulates brainwave rhythm would eventually result in flattened cortical brainwaves. This may imply that cortical brainwaves originated from deep structures inside the brain (the larger limbic system), and that brain activities did indeed start deep within the brain's center, involving anatomical locations that have extensive networks with the cortices.

In the case of cardiac death, the cardiopulmonary system ceases to function, and the brain ceases to function within a few minutes (3–5 minutes). When most people consider the definition of death, they envision this form of demise. As a result, in cases of cardiac death, the person's pupils are frequently noticed as static and enlarged, and so are the vital signs.

(Wavy items like heart rate, blood pressure, and respiration are missing.) As a result, what appears to be cardiac death is tied to brain death as well.

All of this implies that the brain is superior to the heart, and that the seat (center) of the human soul most likely resides in the brain at the greater limbic area; it may not be the observable anatomical structures in this area per se, but rather an 'unseen' element at this particular deep entered anatomical area (it is worth noting that the initial historical discussions on humans' seat of the soul and the greater limbic system are primarily intended for death status and distinctive humor).

In summary, the following 5 points should be highlighted:

(A) The brain appears to be superior to the heart for the following reasons:

  i) The state of a person's brain function is the most crucial in deciding death;

  ii) vital signals of the cardiopulmonary system, such as heart rate, blood pressure, and breathing (wavy things), can be sustained by a machine and pharmaceuticals;

  (iii) flattened brainwaves look unlikely to be reversible to wavy brainwaves in a dead person, and no mechanism may be capable of causing the emergence of 'permanent wavy' brainwaves in a dead person,

B) Waves (ups and downs, ups and downs, right-left, left-right oscillations) may be 'indirect' expressions of the soul; once dead, all waves flatten and eventually all atoms stop vibrating (non - wavy), and the physical dimension begins to disintegrate. Keep in mind that atoms can act as either particles or waves. The phenomenon is known as wave particle duality for an atom,

C) Brainwaves can be regarded to 'visualize thought' as 'images'; thus, more studies are needed to correlate brainwaves with brain anatomy, and, indeed, advanced technology is clearly required to enable scientists to examine deep brainwaves noninvasively and correlate them with cortical (superficial) brainwaves, brain anatomy and functions,

D) everything is a wave (ups and downs, energy, life, the will to live, an indirect expression of the soul or everything is the soul) and lastly

E) studies on waves, oscillations, frequency and physiology (even anatomy, simply because atoms can also behave as waves) could in fact be studies related to the soul.

2.10. CONCLUSION

To the general public, it is represented by the enormously popular public exhibitions of plastinated cadavers and body parts. For some, the discipline of anatomy is characterized by formalin cadavers in sterile dissecting rooms and very large amounts of detailed anatomy. However, the humanistic side of anatomy reminds us that anatomy is more than merely a science.

Within academia, it has frequently been seen as the mainstay of medical teaching, even as a handmaid of surgery. These movements can be traced through the many editions of Gray's Anatomy, from 1858 to the present day.

The one unitary theme across this broad swathe of biomedical endeavor is structure or organization, the fundamental thread that ties together all these approaches within modern anatomical science. However, these descriptions amount to too little more than facets of what constitutes anatomy in modern guise, with its strong research ethos and broad scope from biological anthropology and clinical anatomy to molecular biology and genetics.

REFERENCES

1. Fallavollita, P., 2017. Innovative Technologies for Medical Education. Human Anatomy - Reviews and Medical Advances, [online] Available at: <https://www.intechopen.com/chapters/55203> [Accessed 23 June 2022].
2. Gangemi, A., Chang, B., Bernante, P. and Poggioli, G., 2021. Robotic Surgery: Rediscovering Human Anatomy. International Journal of Environmental Research and Public Health, [online] 18(23), p.12744. Available at: <http://dx.doi.org/10.3390/ijerph182312744> [Accessed 23 June 2022].
3. Hegazy, A., 2019. Human Anatomy: An Inlet of Medicine and Surgery. International Journal of Human Anatomy, [online] 1(4), pp.1-1. Available at: <http://dx.doi.org/10.14302/issn.2577-2279.ijha-19-2976> [Accessed 23 June 2022].
4. Idris, Z., Reza, F. and Abdullah, J., 2017. Human Brain Anatomy: Prospective, Microgravity, Hemispheric Brain Specialization and Death of a Person. Human Anatomy - Reviews and Medical Advances, [online] Available at: <https://www.intechopen.com/chapters/54586> [Accessed 23 June 2022].
5. Jones, D., 2017. Human Anatomy: A Review of the Science, Ethics and Culture of a Discipline in Transition. Human Anatomy - Reviews and Medical Advances, [online] Available at: <https://www.intechopen.com/chapters/55062> [Accessed 23 June 2022].

CHAPTER 3

ANATOMY OF EXTRA-MUSCULAR SOLEUS VEINS: CLINICAL IMPACT

CONTENTS

In different dysfunctions of this system, the venous system of the lower limbs has great structural and functional anatomical complexity which must be considered. Mainly in the venous return, this complexity lies, from the upright position and ambulation and other factors which is changed such as level of physical activity, ambient temperature, heart function and circulating blood volume. The anatomical description of soleus veins (SV) has received little attention from anatomy texts and books. As the site for deep vein thrombosis (DVT), soleus veins have been implicated. Using noninvasive ultrasound techniques detailed anatomical knowledge is required for early diagnosis. We describe the anatomy of the veins that emerge from the ventral surface of the soleus muscle in the present work. The number of veins per leg ranged from 7 to 38. The distribution of these veins per quadrant ranged from 0 to 12. In the upper lateral quadrant, the greatest number of veins occurred.

3.1. CARDIAC ANATOMY FOR THE ELECTROPHYSIOLOGIST WITH EMPHASIS ON THE LEFT ATRIUM AND PULMONARY VEINS

Thanks to a rigorous understanding of cardiac anatomy, the recent development of catheter ablation was possible. To avoid or minimize complications during catheter placement and RF application, appropriate cardiac structure knowledge is relevant. After a meticulous characterization of the atrial muscular sleeves that prolong inside the veins, new strategies for pulmonary vein isolation appeared and made the procedure safe and efficient.

To understand catheter placement and ablation targets this chapter aims to provide basic anatomical knowledge for interventional electrophysiologists. We begin with the location of the heart inside the mediastinum, position of cardiac chambers, pericardial space, and neighboring structures of the heart. We continue with the right atrium and important structures inside it: Koch's triangle, sinus node, cavotricuspid isthmus, and interatrial septum with fossa ovalis.

3.1.1. General Anatomy of the Heart

Between the two lungs, the heart is positioned 2/3 to the left and 1/3 to the right of the midline of the thorax. The anterior part of the heart consists of the right ventricle, it lies behind the sternum. In front of the spine, the base of the heart lies. By the pericardium, the neighboring structures are separated from the heart.

With the esophagus, the posterior wall of the left atrium comes in contact, which can be close to the right or left orifices of the pulmonary veins. To avoid the risk of atrioesophageal fistula, catheter ablation at this level should be performed with lower energy or with temperature monitoring.

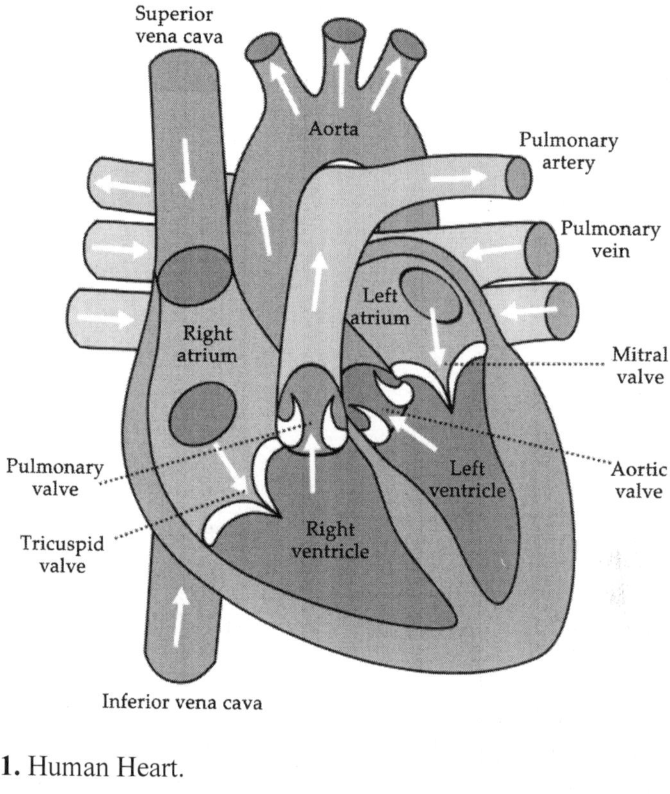

Figure 3.1. Human Heart.

Source: Image by Wikimedia

The right and left phrenic nerves descend on the outer surface of the pericardium. The right nerve can be damaged during cryoablation of the RSPV or RF ablation near the sinus node and is close to the superior vena cava and right superior pulmonary vein (RSPV). At the level of superior vena cava, in order to avoid damage to the right phrenic nerve, the right atrium to distinguish hiccups and avoid ablation lesions at this level and high-output stimulation is performed.

When ablating at the base of the appendage, especially in patients with persistent atrial fibrillation when extensive ablation is needed the left phrenic nerve is in the proximity of the left atrial appendage and can be damaged.

3.1.2. The Right Atrium

The right atrium has four classical components: the interatrial septum, the vestibule, the venous part and the right atrial appendage. At the level of the Cavo atrial junction with superior vena cava, the sinus node is located on the anterolateral part of the right atrium. At this level with a diagnostic catheter, the right atrial appendage has prominent muscular bundles that give the high amplitude potential recorded.

Without pectinate muscles, present in other regions of the right atrium the vestibule surrounds the orifice of the tricuspid valve and has a smooth appearance. The venous component lies between the superior and the inferior vena cava and forms the posterior aspect of the right atrium. By the terminal crest, it has a smooth wall that is separated from the pectinate muscle atrial zone.

An important region between the inferior vena cava and the vestibule of the tricuspid valve that is "burned" during RF ablation of typical atrial flutter the terminal part of the crest divides into small muscular bundles that form the Cavo tricuspid isthmus.

A thin flap that is an important marker for catheter ablation of atrial flutter, the inferior vena cava has a fibrotic partial valve that is called the Eustachian ridge. At the level of the Cavo tricuspid isthmus an important percentage of patient's present pouches and recesses. The ablation is made more difficult by these structures.

The tendon of Todaro is known as the fibrotic prolongation of the Eustachian valve toward the septum. An anatomical structure that every electrophysiologist should know with the septal leaflet of the tricuspid valve and the coronary sinus, orifice forms the triangle of Koch. At this level, sometimes accessory pathways can also be ablated at this level and ablation of intranodal re-entry is performed.

To access the left atrium through transseptal puncture the interatrial septum is used. At this level, the puncture is safe without the risk of pericardial bleeding the true septum that can be crossed with a transseptal needle is the fossa ovalis.

The electrophysiologists are tented to use it to the left atrium when a patent foramen ovale (PFO) is present. The PFO directs the catheter toward the anterior and superior wall of the left atrium, making difficult the ablation of the right pulmonary veins owing to its location. The electrophysiologist should know that the risk of roof perforation is higher when using the PFO.

In different types of arrhythmias, catheters positioned inside the right atrium facilitate the understanding of the activation sequence: in case of a typical atrial flutter, a Halo catheter with 20 pols positioned along the tricuspid valve records the electrical signals of counter-clockwise activation; a circumferential mapping catheter placed at the base of the superior vena cava records the electrical activity when mapping sinus node re-entrant tachycardia.

3.1.3. The Left Atrium

The left atrium (LA) is the most posterior cardiac chamber. The esophagus, the descending thoracic aorta and more posteriorly the vertebral column, behind the LA lies the tracheal bifurcation.

The left atrium is a structure composed of four parts: the vestibule of the mitral valve and the left interatrial septum, the venous component that receives the pulmonary veins, the left atrial appendage. The walls of the left atrium are posterior, anterior, superior, left lateral and septal. To the horizontal plane the interatrial septum has a 45–60 degrees angulation.

The left appendage presents pectinate muscles, whereas the superior and posterior walls of the left atrium are smooth. As described by Biase et al. the left atrial appendage has a particular morphology: cauliflower 3%, cactus-like 30%, chicken wing 48% and windsock 19%. At this level patients with the chicken wing morphology are less likely to develop thrombus. The left appendage is smaller than the right one.

Because the left atrium lies between the ascendant aorta anteriorly the transverse diameter of the left atrium is the largest and the spine posteriorly and the dilation of the cavity is made between these structures. To the right pulmonary artery and the bifurcation of the pulmonary trunk, the roof of the left atrium is close.

There are muscular bridges made of atrial myocardium between the two atrial chambers left and right. Extending from the left appendage to the right appendage, the most important is the Bachmann bundle, which is composed of parallel myocardial strands.

With dilation and fibrosis, atrial fibrillation results in remodeling of the left atrium (the so-called atrial cardiomyopathy). The left atrial isthmus, is the connecting line between the inferior margin of the LIPV and the mitral annulus, which is not a distinct anatomical structure. When ablating persistent atrial fibrillation and increasing the success rate of the technique, the line is used.

3.1.4. The Pulmonary Veins

From the lungs to the posterior aspect of the left atrium the pulmonary veins drain oxygenated blood. Then the right ostia, the left pulmonary vein ostia are located more superiorly. The right and inferior pulmonary veins are posterior and downwards whereas the right and left superior pulmonary veins are anterior and superior structures.

With a super inferior diameter longer than the anteroposterior diameter, the orifices of the veins are oval. Sometimes supplementary veins can be found, more frequently on the right side, but usually, there are two veins on the right and two on the left side. On the left side, a frequent anatomical variation is the presence of a common trunk.

Developing muscular sleeves, the musculature of the left atrium extends inside the pulmonary veins; the longest being found in the superior veins: LSPV and RSPV. The myocardial fibers extend at a length of 1–3 cm. Usually, on the inferior part of the superior veins and on the superior part of the inferior veins the sleeves are more important.

By the left ridge the superior pulmonary vein is separated from the left atrial appendage, during RF ablation of atrial fibrillation which is a structure that needs to be ablated because muscular sleeves are very well developed at this level. The most challenging part is to obtain a good contact with the ridge.

From the vein, there is a direct link between the pressure inside the pulmonary veins and abnormal electrical activations. The junction between the LA and pulmonary veins becomes the source of abnormal activations from the pacemaker cells when the left atrial pressure increases above 10 cm H2O (Cajal-like cells).

Because ectopic triggers are found inside the pulmonary veins the modern treatment of paroxysmal atrial fibrillation is pulmonary vein isolation. With a high risk of pulmonary stenosis ablation of pulmonary foci is effective, at the level of venous antrum and aims to isolate the veins from the atrium therefore in the last years, ablation is performed.

3.1.5. The Right Ventricle

The right ventricle is the heart chamber that is situated the most anterior. It has three portions: the outlet or RVOT that is continued with the pulmonary artery, the inlet which is delimitated by the tricuspid valve and papillary muscles, and the apical part. Inside the right ventricle, a thick moderator band can be present, making catheter manipulation difficult inside the RV.

In a posterior position, the RVOT is superior to the left ventricular outflow tract which crosses the RVOT. When a stiff tip ablation catheter is advanced directly to RVOT the myocardium of RVOT is very thin, and perforations can result.

The RVOT is the most frequent region of benign monomorphic premature ventricular complexes. Ablation is carried out at this level using pace mapping and activation mapping at the level of septal, lateral, anterior, and posterior RVOT. When ablation is not effective in the RVOT, this structure is in close relation with the left ventricular outflow tract and aortic cusps, and sometimes imaging of these structures should be performed.

3.1.6. The Left Ventricle

By the mitral valve the left atrium is continued and the left ventricle also has three components: inlet, outlet and apical part. From the insertion of the papillary muscles to the apex the apical LV extends. The walls of the left ventricle are thicker than those of the right ventricle but the trabeculations are finer than those of RV. From the PVC, arising from the left bundle conduction system, the posteroinferior papillary muscle and the anterosuperior papillary muscle can be sources of ventricular premature contractions that have to be differentiated.

Superiorly and anteriorly the LVOT is directed. It can be a source of ventricular premature contractions. Through the transmittal anterograde approach access can be achieved that requires a transseptal puncture or through the transaortic retrograde approach when mapping the left ventricle.

3.1.7. The Coronary Veins

By the coronary venous system, most of the venous flow of the heart is collected. The coronary sinus drains the great cardiac vein and the middle cardiac vein as well as other small veins. For catheter ablation of ventricular premature contractions or ventricular tachycardias the cardiac veins might be used.

Through the venous system, electrophysiologists can reach the epicardium of the left ventricle. Ablation is performed usually with irrigated-tip catheters to avoid perforation of outer walls that are not protected by muscular bundles and small diagnostic and therapeutic (2F–5F) catheters are used inside the coronary veins.

Sometimes muscular bundles prolong into the left atrium and form sleeves that cover the coronary sinus. With persistent atrial fibrillation that needs substrate modification, these muscular bundles are a target of ablation in patients.

Sometimes the middle cardiac vein forms a diverticulum of the coronary sinus and is dilated. At this level, as poster septal accessory pathways could be located at this level catheter ablation might be performed. By a small flap, the coronary sinus orifice is bordered: by a diagnostic catheter because this valve is incomplete, the Thebesian valve is easily passed.

3.2. ANATOMICAL, BIOLOGICAL, AND SURGICAL FEATURES OF BASAL GANGLIA

Basal ganglia influence the information in the extrapyramidal system and they refer to the deep gray matter masses on the telencephalon and encompass a group of nuclei. With the nervous system in humans, they are related to numerous significant functions controlled. Gross anatomically, they are comprised of different parts as the dorsal striatum that are consisted of the caudate nucleus and putamen and ventral striatum which includes the subthalamic nucleus, the nucleus accumbens, globus pallidus, olfactory tubercle, and substantia nigra.

Nucleus accumbens, is also associated with reward circuits and has two parts; the nucleus accumbens shell and the nucleus accumbens core. There are important findings explaining striatal neurodegeneration in the human brain: neurological diseases are characterized by the obvious pathology of the basal ganglia.

By bacterial and/or viral infections some of these diseases are induced. For neuronal disease treatment like Parkinson's Disease or Thiamine Responsive Basal Ganglia Disease or Wilson's Disease, respectively surgical interference can be one alternative in addition to vascular or tumor surgery within this area. To the optimization of the diagnosis and later patient's treatment extensive knowledge on the morphological basis of diseases of the basal ganglia along with motor, behavioral and cognitive symptoms can contribute significantly.

3.2.1. Structure and Function of Human Basal Ganglia

The deep gray matter masses on the deep telencephalon and encompasses a group of nuclei known as basal ganglia. Generally, in the extrapyramidal system and in human beings, basal ganglia influence information and are related to numerous significant functions controlled by the nervous system, including control of habitual behavior, voluntary motor movements, procedural learning, such as eye movements, cognition, and emotions. The nuclei group is placed deep beneath the cortical area of the brain.

Additionally, the basal ganglia specialize in processing data on movement and in fine adjustment of the brain circuit activity that in specific habitual conditions/actions define the best suitable response such as riding a bicycle, playing a piano, and so on. While planning movement and learning novel actions in new situations, they also play a major role.

The central nervous system (CNS) develops in the early stages of embryological development in many mammals, from an embryological point of view, including humans, from the neural plate. To form a neural groove and then differentiate into the neural tube in humans, the ectodermal cells from epiblast fold in the middle of the third week of development.

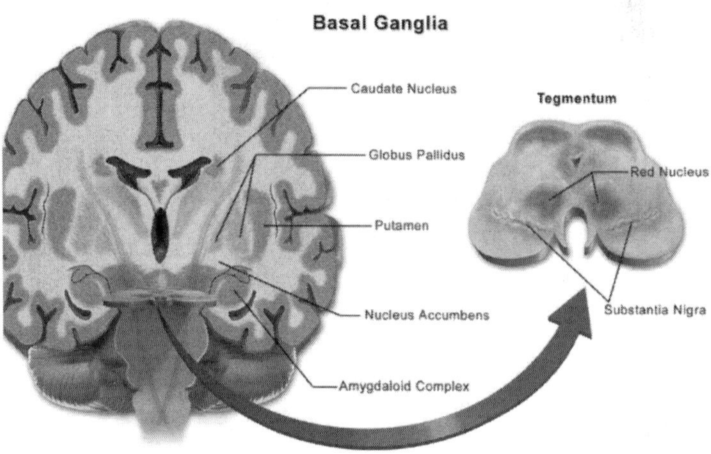

Figure 3.2. Basal Ganglia.

Source: Image by Wikimedia

The anterior neuropore, on the 24th day, and then the posterior neuropore, on the 27th day are closed, while the cerebrospinal fluid is secreted by the ependymal cells, which are lining within the ventricular system, pros encephalic, mesencephalic, and rhomb encephalic vesicles develop the spinal cord from the caudal part of the neural tube and from the rostral part of the neural tube.

From a part of the pro's encephalic vesicle at the late 34th day, while telencephalic vesicles developed, anlage primordium of the future cerebral cortex, basal ganglia, and olfactory bulb can be clearly identified at the 36th day. Near the other epithelial structures such as thalamic, hippocampal, and hypothalamic epitheliums, at the eighth week of the development of the basal ganglia, the neuroepithelium is clearly defined.

From this epithelium, basal ganglia neurons are derived. The neuroepithelial cell layer becomes thicker and new neurons are generated between the 19 and 22 gestational weeks, which migrate into the striatum. In basal ganglia, between the 23 and 28 gestational weeks, the neuroepithelial cell layer over some places of the caudate nucleus became faint.

Again, neuroepithelial cellularity is scant at week 27 of the development in the basal ganglia. In the basal ganglia, between the 29 and 33 gestational weeks, neuroepithelium was thicker over the striatum and nucleus accumbens. The number of the glial cells increase in the 32nd week of the development process. During this period, the onset of myelinization is very silent on the other hand.

It can be detected microscopically in the subventricular zone, at the 35th week, the glial cell number is increased and neuroepithelial cells are limited. In the internal capsule, there is a slight onset of myelinization. While the numbers of astrocytes decrease in the internal capsule and basal ganglia, at term, proliferating neuroepithelial cells can be detected in the subventricular zone.

In the postnatal period, myelinization of the internal capsule is complete approximately at the 6th week postnatally and basal ganglia and diencephalic neurons were well organized. In basal ganglia, two years postnatal, myelinization of the internal capsule appear completely and all the mature histological structures can be clearly detected in the caudate and putamen.

It is aimed to highlight the anatomical, biological, and surgical importance of cortical-basal ganglia circuits and their role in the pathogenesis of neurological processes in our investigation. An opinion is developed that detailed anatomy related information in embryology, histology, and gross anatomy as well as molecular and surgical information, depending on the facts available nowadays and our experience, after intervention to the region of intervention, on the basal ganglia and neighborhood structures may cause confused clinical outcomes and possibly the option of renovating the morphological brain structure.

Including the dorsal striatum consisting of the caudate nucleus and putamen and ventral striatum, gross anatomically, the basal ganglia are composed of different parts, which include the subthalamic nucleus, the nucleus accumbent, globus pallidus, substantia nigra, and olfactory tubercle. Defined, complex internal morphological and biological features are shown by each of these parts.

On each of the cerebral hemispheres, basal ganglia are composed of several subcortical nucleus groups that are located deep. They are called "lentiform nucleus," which includes both the putamen and globus pallidus; the "striatum," which includes the nucleus caudates and putamen; and "corpus striatum," including the caudate nucleus and the nucleus lentiform is, and the others including their projections, subthalamic nucleus, claustrum, and nucleus accumbent.

The substania nigra is placed on the mesencephalon which is also a basal nucleus. This nuclear group carries phylogenetically, functionally, terminologically, and heterogeneous formations. The amygdaloid body was a part of the basal ganglia (archistriatum) according to the previous classification. Anatomists considered it as a functional educative part due to the acquisition of new scientific data associated with it.

On the lateral to the putamen and medial to the insula clastrum is located, which was noted by some sources to be a part of the basal ganglia. As a simple relay within the basal ganglia, the globus pallidus externus was originally discovered.

In the basal forebrain region superior to the preoptic region of the hypothalamus the nucleus accumbent is also associated with reward circuits that are located, while the prefrontal area was on both cerebral hemispheres, whose mission is planning and motivating movement performed by the body. The nucleus accumbent has two parts: the nucleus accumbens shell and the nucleus accumbens core and, which consist of their own morphology and functions.

Basal ganglia circuits contain an assortment of cell types that mediate synaptic interactions within and between basal ganglia nuclei. In the different areas of the basal ganglia neurotransmitters also play an important role. For example, dopamine, is the source of the striatal input in the substantia nigra which has a very important function within the basal ganglia.

3.2.2. Pathogenic Conditions of the Basal Ganglia

In the substantia nigra, Huntington's and Parkinson's diseases are caused by the degeneration of dopamine-producing cells. On the other hand, which possesses an inhibitory effect on the target neurons, most of the neurons in the basal ganglia use gamma-Aminobutyric acid (GABA) as a neurotransmitter. By both external inputs to the striatum and by a group of striatal neurons, acetylcholine is another important neurotransmitter and is regularly used. One of the major acetylcholine concentration regions is the striatum although the total number of cholinergic neurons is the smallest in all brain neurons.

The striatum arises from numerous large and small bundles of nerve fibers and is currently considered to be the largest region of the basal ganglia. The histological organization of the striatum is considered very complex. The major populations of striatal neurons are medium spiny neurons, densely covered dendritic spines that receive input from the cortex and the thalamus, which are GABAergic cells with small bodies. In the striatum, with smooth dendrites the cholinergic neuron population comprises the cholinergic neurons.

Both the globus pallidus and the ventral pallidum consisted of the pallidum. Into the internal and external segments, the globus pallidus can be divided functionally. Histologically, both segments have primarily GABAergic neurons. The internal segment receives input from direct and indirect pathways, while external segments receive inputs mainly from the striatum and pass through the subthalamic nucleus. A mechanism in which there are inhibitory effects on the target, pallidal neurons function basically via de-inhibition.

The substantia nigra is a mesencephalic gray matter portion of the basal ganglia, which is divided into two parts: pars compacta and pars reticulate. Although pars compacta produce dopamine, which plays a major role as a regulator neurotransmitter in the striatal pathway, the pars reticulate has inhibitory effects on the thalamus.

The subthalamic nucleus is a diencephalic gray matter portion of the basal ganglia and produces glutamate, which is an excitatory neurotransmitter in the ganglia. This nucleus, also sends excitatory input to the internal part of the globus pallidus while receiving inhibitory input from the external part of the globus pallidus.

3.2.3. Basal Ganglia-Related Pathological Conditions

There are important findings that explain striatal neurodegeneration on the human brain clinically, many neurological diseases are characterized through the obvious pathology of the basal ganglia. By bacterial and/or viral infections some of these diseases are induced where the bacteria and/or virus introduces some genetic material and, as a consequence, by the cytokine regulation, either activated or downregulated some of the life's essential processes or other diseases that are affected in association with neurodegenerative diseases like TGF-β or TNF-α related to medical solutions like in dentistry in addition to applications.

Here, and as a consequence, affecting either the neuronal structure or the functionality or both of them can be seriously affected. Lately, more knowledge about the problems of the basal ganglia patients with neurodegenerative, neuropsychiatric diseases, inflammatory, immunologic, vascular, metabolic, allergic, congenital, traumatic, endocrine and malignant became available.

To the available knowledge related to the pathogenesis of neurological diseases a comprehensive understanding of the striatal projection loss while receiving striatal input/output on the neurons will contribute. As it is the case for Parkinson's disease, surgical intervention can also be one alternative for neuronal disease treatment, thiamine responsive basal ganglia disease or Wilson's disease, respectively, within this area in addition to vascular or tumor surgery.

Tremors, grimaces, and repetitive movements are caused by the lesions of the basal ganglia. In different pathological processes, at the same time, such as Kernicterus, Tourette syndrome, hemiballismus, obsessive-compulsive disorder, neonatal and lacunar infarction, Huntington's and Parkinson's diseases, basal ganglia neurons were affected. Again, in carbon monoxide poisoning, selective necrosis is caused in the globus palidus.

3.2.4. Potential Approaches for Basal Ganglia Disease Treatment

Gene therapy-based treatment modalities bear potential for the treatment of nervous system diseases or disorders, in order to limit or inhibit this type of disorders, these include viral vector systems, gene-based vaccines and immunotherapy, anti-nervous system degenerative diseases treatment by molecular regulators RNAi applications, plasmid DNA applications, cytokine targeting like TNF-α targeting, epigenetic targeting.

Several studies on basal ganglia supported by data aligned to age- and/or gender-dependent relation of intelligence with volumes of the nuclei were presented recently, gender distinctive diseases on the subcortical nuclei and still limited results are known regarding the potential influence of age. The basal ganglia (BG), are significant for connection among forebrain nuclei, which play a major role in selecting and shaping motor and cognitive behaviors.

Surgically, using deep brain stimulation (DBS), which is an implanted electrical device modulating distinct targets at the brain, in some neurologic diseases, resulted in the symptomatic development of movement disorder, especially in both hyper- and hypokinetic movement disorders of the basal ganglia deep brain stimulation (DBS) that is considered highly effective. For the disorders of these regions, the clinical benefit of DBS is based on the experience with prior surgical ablative therapies, and, in part, used by neurosurgeons decades ago.

By electrical or radiofrequency lesioning of that region before DBS are treated, the most commonly DBS-treated conditions were and are Parkinson's disease and dystonia. For both procedures, the duration and temperature are important. Nervous system disease or injury can be restored by applying electrical current to the functions that were partially or totally lost.

Some entry points described by the authors for nucleus accumbent were as follows, in stereotaxic surgery: 4–10 mm lateral to the median line, 19–23 mm prior to the midpoint, and 7–9 mm below the anterior commissure-posterior commissure (AC-PC) line. The original target is the core of the nucleus accumbens. For deep brain stimulation applications as refractory major depression, Tourette syndrome and obsessive-compulsive disorders the stereotactic coordinates were as follows: 6.5–8 mm lateral to the midline, 4–4.5 mm ventral to the AC-PC plane and 1.5–2.5 mm anterior to the anterior border of the AC.

With accompanying major depression in treating obsessive-compulsive disorders, as the tips of the electrodes Aouizerate et al. reported their experience in DBS targeting, were situated 31.4 mm anterior to the PC on the left side, 3.0 mm below the AC-PC line, 8.9 mm lateral to the AC-PC line, and 36.5 mm anterior to the PC on the right side, and 1.7 mm below and 7.6 mm lateral to the AC-PC line.

The subthalamic nucleus (STN) of advanced Parkinson's disease patients undergoing deep brain stimulation application is a prominent target for treatment. To identify a significant target, microelectrode recording (MER) is used in many patients. As a training set and found the error in predicting the STN entry to be (mean ± SD) 0.18 ± 0.84, and 0.50 ± 0.59 mm for the STN exit point, moran in a previous work showed that trajectories served, which yields a 0.30 ± 0.28 mm deviation by using MER from the expert's target center.

There was a negative correlation between right substancia nigra (SN) volume and the unified Parkinson disease rating scale (UPDRS) score ($r = -0.466$, $p = 0.038$) in the correlation analysis, and there was a tendency but correlation between the left SN volume and UPDRS score ($r = -0.443$, $p = 0.050$) is not significant.

It was shown in a previous approach, between both the left and right SN that subjects suffering from Parkinson's disease showed a significant asymmetry, nucleus caudatus, and nucleus lentiformis volumes ($p = 0.001$, $p < 0.001$, $p = 0.044$), between the volumes of left and right SN, with taking into account that the control subjects also showed a significant asymmetry, nucleus caudatus, and nucleus lentiformis ($p < 0.001$, $p = 0.003$, $p < 0.001$, respectively).

It was found that the left basal ganglia structure was smaller than the left ones, in addition, during our experience using stereological methods on the basal ganglia volumetry on right-handed patients with Parkinson's disease. However, only substantia nigra possessed a smaller volume, when we compared them with the control cases.

Also, evaluation of the basal ganglia and substantia nigra volume in Parkinson's disease (PD) in comparison to the healthy age-matched control subjects patients revealed significant atrophy in SN. However, during the study. Basal ganglia and the SN are the regions with predominantly pathological changes in PD significant atrophy in nucleus lentiformis and nucleus caudatus was not found.

However, there is no study to our knowledge in the literature that evaluates the asymmetrical volume changes by using the stereological technique, there are studies in the literature that examine the volumetric differences in basal ganglia and SN anatomy in PD. By overlaying each selected section using a regular grid of test points that is randomly positioned, Cavalieri's principle of stereological approaches through point counting is accomplished.

For volume estimation in MRI the Cavalieri theorem of systematic sampling combined with point counting proved to be a reliable, simple, inexpensive, and efficient method, and during the morphological changes evaluated during Parkinson's disease development this stereological approach can provide valuable information.

For this purpose, can be variable surgical procedures applied. Through a single frontal entry point for six patients Gallina et al. provided details of the surgical procedure for both caudate and putaminal tracks, and for the following 10 procedures and they used two completely distinct routes, with two separate entry points, each for the nucleus caudatus and putamen, respectively.

To ensure optimal positioning of desired targets in surgical processes, surgeons use a stereotactic frame that helps, or frameless stereotactic systems or, alternatively, neuronavigation or electrophysiological mapping of the brain for lesioning-related basal ganglia and to protect the surrounding neural tissue for obtaining the main target in a three-dimensional manner in addition.

3.3. MESENCEPHALON; MIDBRAIN

The mesencephalon sits above the pons and is adjoined rostrally to the thalamus and the most rostral part of the brainstem. It comprises two lateral halves, called the cerebral peduncles; which is again divided into a posterior part, tegmentum and an anterior part, the crus cerebri. The tectum is dorsal to an oblique coronal plane which includes the aquaduct, and consist of pretectal area and the corpora quadrigemina. The cerebral peduncles are seen to be composed of dorsal and ventral regions separated by the substantia nigra in the transverse section.

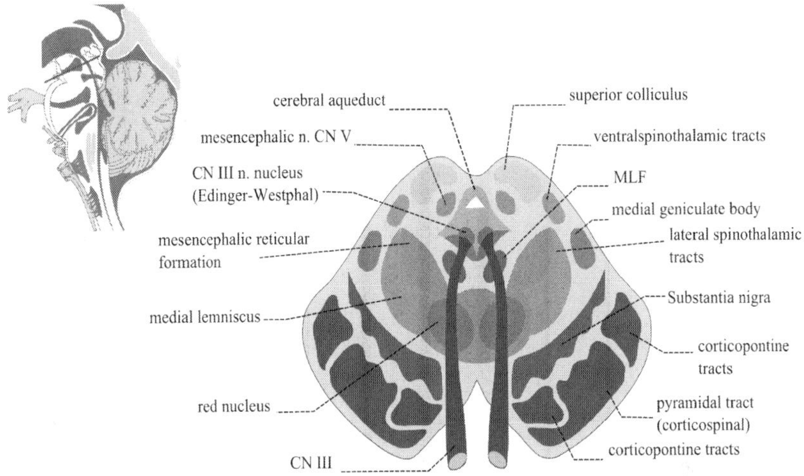

Figure 3.3. Parts of midbrain.

Source: Image by pixabay

Tegmentum mesencephali contains medial longitudinal fasciculus, oculomotor nucleus, red nucleus, thochlear nucleus, reticular nuclei, lateral lemnisci and medial lemnisci. In the tectum, which are continuous with the periaqueductal grey matter, the inferior colliculus and superior colliculus have the main nucleus. Particularly in movements of the eye, and in auditory and visual processing the mesencephalon serves important functions in motor movement.

The mesencephalic syndrome causes tremor, depression or coma, spastic paresis or paralysis, nystagmus and opisthotonos. In addition, brain tumors, inflammatory, thiamin deficiency and cranial trauma or degenerative disorders of the mesencephalon have also been associated with the midbrain syndrome.

The central nervous system and the peripheral nervous system are two components of the nervous system. Brain and spinal cord constitute the central nervous system. The peripheral nervous system consists of nerves, sensory neurons and ganglia which are connected with each other and with the central nervous system. The brain is a component of the central nervous system. It contains three basic subdivisions, namely the cerebellum, cerebral hemispheres and brainstem.

3.4. EMBRYOLOGY

From rostral to caudal, embryologically, the central nervous system can be divided into five continuous parts, they are:

- The cerebrum (telencephalon) becomes two cerebral hemispheres. The hemispheres are partially separated by a deep longitudinal fissure and the surface of these hemispheres consists of gyri and sulci.
- By the cerebral hemispheres, the diencephalon is hidden from view in the adult brain. The thalamus, hypothalamus, and other related structures consisted by it and classically is the most rostral part of the brainstem.
- The mesencephalon (midbrain), located at the junction between and in both middle and posterior cranial fossae is the first part of the brainstem seen when an intact adult brain is examined.
- The cerebellum and the pons by the metencephalon.
- The myelencephalon (medulla oblongata), ends at the foramen magnum, the caudal most part of the brainstem.

In the region where the earliest somites appear closure of the neural tube first occurs; closure spreads cranially and caudally. The cranial and caudal neuropores are the unfused regions of the neural tube. Some fundamental subdivisions in the early nervous system have become manifest, even before the closure of the neuropores (24 days of gestation for the cranial neuropore, and 26 days of gestation for the caudal neuropore).

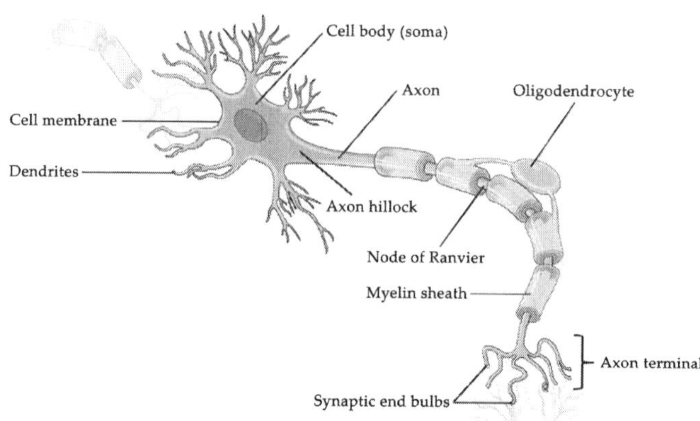

Figure 3.4. Parts of a neuron in the CNS.

Source: Image by Flickr

Within the brain the forebrain (prosencephalon), midbrain (mesencephalon), and hindbrain (rhombencephalon) can be distinguished, the future spinal cord and brain are recognizable. The overall bending of the cephalic end of the embryo into a "C" shape, a prominent force in shaping the early nervous system. At the level of the mesencephalon at the end of the third week associated with this bending is the appearance of a prominent cephalic flexure of the brain.

A cervical flexure appears at the boundary between the hindbrain and the spinal cord at the beginning of the fifth week. The original three-part brain has become subdivided further into five parts by week 5. The mesencephalon remains undivided and tubular in its overall structure, which is sharply bent by the cephalic flexure.

3.5. ANATOMY

3.5.1. External Features of the Midbrain

The midbrain is the short part, which connects the pons and cerebellum with the thalamus and cerebral hemispheres, constricted portion. It is not more than 2 cm in length, and most of it lies in the posterior cranial fossa, it is the smallest part of the brainstem.

Separated by the interpeduncular fossa, on the anterior surface of the midbrain are located the cerebral peduncles. On each side of the midline, the cerebral peduncle is two large bundles of fibers. By a pigmented band of gray matter called substantia nigra, in addition, it is again divided into an anterior part, *the crus cerebri*, and a posterior part, tegmentum mesencephali.

From the cerebral hemispheres, the crura cerebri are superficially corrugated and emerge. As they enter the pons, they converge as they descend and meet, where they form the caudolateral boundaries of the interpeduncular fossa. By the cerebral aqueduct that connects the third and fourth ventricles, two crura are separate, whereas the tegmental parts are united and traversed.

From the medial aspect of the cerebral peduncle of the same side the oculomotor nerve (CN III) emerges. Exiting from the interpeduncular fossa near the junction of the pons and midbrain are the oculomotor nerves. Two of the extraocular muscles, this cranial nerve supplies all. The interpeduncular fossa is the crus cerebri that contains a midline depression.

At the level of the tentorial incisura, the basilar artery divides in the interpeduncular fossa into right and left posterior cerebral arteries. Around the ventral (basilar) crural surfaces, the superior cerebellar and posterior cerebral arteries run laterally. Between two arteries the trochlear and oculomotor nerves lie. Numerous small holes, in the depths of the interpeduncular fossa can be seen. The entry point of the posterior cerebral artery is represented by these holes. This region is usually referred to as the posterior perforated substance, because of its appearance.

The inferior and superior colliculi one on each side, on the posterior surface of the midbrain are four prominent rounded elevations. Collectively, four colliculi are called as tectum (corpora quadrigemina). By a cruciform sulcus, the superior and inferior colliculi are separated. Into a depression for the pineal gland the upper limit of the sulcus expands.

Over the superior medullary velum, median frenulum veli is prolonged from its caudal end down. The superior colliculi are associated with visual responses, and are larger and darker than the inferior colliculi. The inferior colliculi are associated with auditory pathways and are smaller. Each colliculus coming from respective colliculi, is laterally related to ridges called superior brachium and inferior brachium. To lateral geniculate body superior brachium connects the superior colliculus.

The inferior colliculus is connected by the inferior brachium to the medial geniculate body. Caudal to the inferior colliculi and pass inferiorly around the lateral side of the midbrain, the trochlear nerves (CN IV) arise from the dorsal midbrain. From the dorsal surface of the brainstem, the trochlear nerve is the only cranial nerve that exits. In motor movement, particularly movements of the eye, and in auditory and visual processing, the midbrain serves important functions.

3.5.2. Internal Structure of the Midbrain

By substantia nigra, on the transverse section, the cerebral peduncles are seen to be composed of dorsal and ventral regions separated. The dorsal region is the tegmentum, and the ventral part is the crus cerebri, on each side. The major pathways of motor neurons out of the cortex are cerebral peduncles. Between the substantia nigra and the aquaductus mesencephali is the tegmentum. In the medulla and pons, it also refers to the corresponding regions. Tectummes encephali, contains two superior colliculi and two inferior colliculi, and are located dorsally to the aquaductus mesencephali.

Crura cerebri (the cerebral peduncles, pedunculus cerebri)

A massive band of descending corticofugal fibers, the crus cerebri is the most ventral part of the midbrain. In section, each crus cerebri is semilunar. It contains corticonuclear, corticospinal and corticopontine fibers. In the primary motor cortex, corticonuclear fibers (the corticobulbar fibers) originate. They descend down to the midbrain and through the genu of the internal capsule. The middle third of the crus cerebri contains the corticobulbar and corticospinal fibers, in the midbrain.

In the motor nuclei of the cranial nerves and other brainstem nuclei, the corticobulbar fibers end. The corticospinal fibers travel through the posterior limb of the internal capsule and are white matter motor pathways starting at the cortex. From the pons and then to the medulla, they enter the cerebral peduncle at the base of the midbrain, then pass through the brainstem.

In the spinal cord for direct muscle control, corticospinal neurons synapse directly onto alpha motor neurons. In the cerebral cortex they form two groups corticopontine fibers arise, both of which end in pontine nuclei. From the cells of the frontal lobe, the frontopontine fibers arise, and end in the nuclei of the pons, the temporopontine fibers, traverse the internal capsule, which are largely from the posterior region of the temporal lobe but occupy the lateral sixth of the ipsilateral crus.

***Substantia nigra*:** In the midbrain, the pigmented substantia nigra, the largest single nuclear mass, is connected massively with the basal ganglia, it is considered to subserve a motor function but it has other projections as well. In unstained brain tissue, it looks like a darkened streak; this is where it gets its name, which is Latin for "black substance."

The substantia nigra is actually made up of two anatomically and functionally distinct portions, although it is often referred to as one structure: the substantia nigra pars compacta and the substantia nigra pars reticulata. Neurons in the pars compacta are much more densely packed together (or compact) than those in the pars reticulata. The the compact zone appears as an irregular band of closely packed pyramidal cells or large polygonal containing granules of melanin pigment.

The reticular zone, also known as the stratum intermedium, is composed of irregular-shaped scattered cells that are rich in iron, but they do not contain melanin pigment, and lies close to the crus cerebri. Most of the dopamine neurons of the brain are found in either the substantia nigra or the ventral tegmental area and originate in the midbrain, which is located adjacent to the substantia nigra.

However, these dopamine neurons are found predominantly in the substantia nigra pars compacta. By GABA neurons, the pars reticulata is instead populated largely. Lesions of the substantia nigra or dopamine deficiencies cause Parkinson's disease. Although what exactly causes neurodegeneration in Parkinson's disease it is still not clear, the individual will likely start to experience movement-related problems, when a significant number of these neurons have died, such as tremor, rigidity, slowing of movements, and postural instability—all hallmark symptoms of Parkinson's disease.

Mainly from the caudate nucleus and the putamen, afferent fibers to the substantia nigra arise. To the striatum and certain thalamic nuclei, the efferent fibers of the substantia nigra project.

Mesencephalic tegmentum

Between the substantia nigra and the cerebral aqueduct is mesencephalic tegmentum. It usually contains the reticular formation nuclei, ascending fiber tracts and cranial nerve nuclei.

Tectum

Below the diencephalon, the tectum is located in the dorsal part of the midbrain. The name comes from the Latin word for "roof." The tectum is composed of a set of colliculi superior and colliculi inferior, which resemble small lumps and is reponsible for the initial processing of sensory information from the eyes and ears. The tectum is responsible for visual and auditory reflexes.

***Superior colliculi*:** To the rostral bump on the lateral side of the midbrain the superior colliculus is present. Including the temporal cortices, retina, inferior colliculus and occipital and spinal cord, from a number of sources, it receives afferents. Collicular efferents pass to the spinal cord, retina, lateral geniculate nucleus, pretectum, thalamus and parabigeminal nucleus.

Through the pulvinar relay to primary and secondary visual cortices they pass. In the superior colliculi, the tectobulbar and tectospinal tracts start from neurons and sweep ventrally around the central gray matter to decussate ventral to the oculomotor nuclei and as part of the dorsal tegmental decussations (of Meynert) medial longitudinal fasciculi. To a visual role alone the superior colliculus is not restricted. It also helps the orientation of the eyes and head.

In the direction of the spinal cord region, part of the colliculus sticks out. To respond to different sensory stimuli this key projection helps the brain. With the detection of the direction of movement of an object in the visual fields the superior colliculus is concerned primarily, and in this way, it facilitates tracking, visual orientation, and searching.

The collicular stimulation produces contralateral head movement as well as movements involving the limbs, eyes, and trunk, which implicates the superior colliculus in complex integration between vision and widespread body activity.

Inferior colliculi: As a main auditory (sound) center for the body the inferior colliculus is a part of the midbrain. By white fibers derived from the lateral lemniscus it consists of a compact nucleus of gray matter containing large and small multipolar nerve cells, and more or less completely surrounded. Its primary roles are pitch discrimination, signal integration and frequency recognition. For Auditory Pathway the inferior colliculi are the relay station. It receives fibers from the auditory cortex, the lateral lemniscus, the ipsilateral medial geniculate body and the opposite inferior colliculus. To the ipsilateral medial geniculate body most efferent fibers travel via the inferior brachium. In the geniculate body, some colliculogeniculate fibers do not relay, but continue, with those that do, via the auditory radiation to the auditory cortex area. To the opposite inferior colliculus and the superior colliculus in few, they pass.

Lesions of either the inferior colliculus or its brachium produce defects in auditory reflex, in tonal discrimination, sound localization and in experimental animals. In humans, the effects of such lesions are poorly documented.

3.6. BLOOD SUPPLY OF THE MIDBRAIN

From two sources the brain receives blood: the vertebral arteries and the internal carotid arteries. At the point in the neck, the internal carotid arteries arise where the common carotid arteries bifurcate. Two major cerebral arteries (terminal branches) it branches to form: the anterior and middle cerebral arteries. From the subclavian artery, the vertebral arteries (right and left) arise.

At the level of the pons to form the midline basilar artery they come together on the ventral surface of the brainstem. In an arterial ring called the circle of Willis, the basilar artery joins the blood supply from the internal carotids. Conjoining two major sources of cerebral vascular supply via the circle of Willis presumably improves the chances of any region of the brain continuing to receive blood if one of the major arteries becomes occluded.

Forming two posterior cerebral arteries, the basilar artery undergoes bifurcation at the site of the midbrain. Around the cerebral peduncles both posterior cerebral arteries travel, and branch into the midbrain forming a series of slender, long penetrating arteries for supplying blood to the thalamus and hypothalamus .

From branches of the basilar artery, the mesencephalon receives its blood supply principally, although branches of the internal carotid also contribute. The main vessels supplying this portion of the brainstem include:

- The posterior cerebral artery (terminal branch of the basilar artery)
- The superior cerebellar artery (branch of the basilar artery)
- Branches of the posterior communicating artery (branch of the internal carotid artery)
- Branches of the anterior choroidal artery (branch of the internal carotid artery)

Numerous veins of the mesencephalon arise from capillaries and, in general, run near the arteries. By the basal veins which drain into either the great cerebral vein (Galen) or the internal cerebral veins these veins from an extensive peripheral plexus in the pia mater and are collected.

3.7. CONCLUSION

The anatomy described was reported by Kwakye as a general systematization of the SV. Uhl and Gillot described an overall systematization of the anatomy of the SV more recently, among other authors. These veins, the SV, are clearly visible inside the muscle for those authors, each part being divided into central, close to the septum and peripheral, located laterally. The medial veins of the soleus are oriented horizontally in the peripheral part of the muscle and vertically in the central part and are smaller than the lateral veins of the soleus. At the proximal part of the muscle to connect the fibular veins more laterally these vertical and central veins join the midline. The lateral view directed vertically shows the large volume of the lateral veins of the soleus. In the fibular veins, they join in several trunks ending, above the arcade of the long flexor of the hallux muscle. This explains why above this section, the fibular veins are much larger. Below, they contain the fibrous, inextensible fibular canal. Due to the arrival of those large lateral veins of the soleus above, they are dilated.

REFERENCES

1. Aragão, J., Reis, F., Aragão, F., Aragão, I. and Leão, S., 2017. Anatomy of Extramuscular Soleus Veins: Clinical Impact. Human Anatomy - Reviews and Medical Advances, [online] Available at: <https://www.intechopen.com/chapters/55254> [Accessed 22 June 2022].
2. Cismaru, G., Muresan, L., Mihai, P., Rosu, R., Gusetu, G., Cismaru, A., Pop, D. and Zdrenghea, D., 2017. Cardiac Anatomy for the Electrophysiologist with Emphasis on the Left Atrium and Pulmonary Veins. Human Anatomy - Reviews and Medical Advances, [online] Available at: <https://www.intechopen.com/chapters/55721> [Accessed 22 June 2022].
3. Kurkcuoglu, A., 2017. Mesencephalon; Midbrain. Human Anatomy - Reviews and Medical Advances, [online] Available at: <https://www.intechopen.com/chapters/55330> [Accessed 22 June 2022].
4. Mas, N., Said, H., Tosun, M., Yonguc, N., Soysal, Y. and Karabekir, H., 2017. Anatomical, Biological, and Surgical Features of Basal Ganglia. Human Anatomy - Reviews and Medical Advances, [online] Available at: <https://www.intechopen.com/chapters/55329> [Accessed 22 June 2022].

CHAPTER 4

MALE REPRODUCTIVE ANATOMY

CONTENTS

The male reproductive system consists of several sex organs that play a role in the process of human reproduction. These organs are located on the outside of the body and within the pelvis. The chapter initially talks about the fundamental components of the male reproductive system. Later, the effect of heat stress on the male reproductive system is also discussed. Towards the end of the chapter, a few methods of sperm selection for IVF and epigenetics in male fertility are also discussed.

4.1. INTRODUCTION

Reproduction refers to the production of new offspring, also known as breeding in animals. It includes a set of physiological processes (usually) that take place in the female reproductive system with the association of behaviors and anatomical structures that are necessary to ensure the birth of the next generation of human, domestic, wild, and laboratory vertebrate organisms.

Although these processes take place within the female's system, it is as a means of the fusion of haploid gametes each from the male (sperm cell) and female (ovum) termed, fertilization in vertebrates. Testes, ductus deferens, epididymis, accessory glands, and penis make up the male reproductive system. The males' reproductive system functions mainly in the production, nourishment, and temporary storage of male gametes (spermatozoa), which are produced via spermatogenesis.

It produces androgens and estrogens through steroidogenesis and very importantly, is connected to the organ of copulation (penis) which serves to introduce semen containing spermatozoa into the female genital system via mating.

4.1.1. Embryology of the Reproductive System

The primordial germ cells have moved from their embryonic stage to the gonadal edges by a month and a half of development in the two genders, where they are encircled by the sex chromosomes to form testes. The forming testicle, whether chromosomally XX or XY, is potential until this point.

The current theory is that the development of an ovary or testis is determined by the synchronized action of a series of genes that contribute to the development of the ovary when there is no Y chromosome or there is no Y testicular development. Unless a gene on the shooting arm of the Y named

TDF (testis defined factor) acts as a switch, the ovarian pathway is followed, diverting development into the male pathway..

One of the main worries in clinical hereditary genes is the quest for the fundamental testis determining factor. The medullary tissue structures customary testicles with seminiferous tubules and Leydigbrates cells in the presence of the Y chromosome that become equipped for androgen secretion under the stimulation of human chorionic gonadotropin (HCG) from the placenta.

Spermatogonia, created by at least 200 progressive mitoses from the early-stage germ cells, forms the walls of the seminiferous tubules alongside the supporting cells. The testicle, of course, delivers an ovary assuming no Y chromosome is available; the cortex forms, the medulla relapses, and oogonia begin to create inside follicles. Oogonia are gotten from germ cells by a grouping of roughly 30 mitoses, not exactly the number essential for spermatogenesis.

Oogonia join meiosis 1 at about the finish of the third month, however, this cycle is interfered with at a point called dictyotene, in which the cell continues until ovulation happens quite a long while later. A considerable lot of the oogonia degenerate before birth, and during the 30 years or so of sexual development of the female, something like 400 matures into ova.

Thickenings in the edges propose the creating of genital channels, the mesonephric (previously called Wolffian) and paramesonephric (previously called Mullerian) ducts, while the early-stage germ cells are relocating to the genital edges. In the male, androgen is delivered by the Leydig cells of the fetal testicles, which stimulates the mesonephric pipes to form the male genital ducts, and Sertoli cells produce a chemical that blocks paramesonephric channel development.

The mesonephric ducts relapse in the female (or in the non-monadic undeveloped organism) and the paramesonephric channels form into the female pipe system. The external privates comprise a genital tubercle, matched labio scrotal swellings and matched urethra folds in the early undeveloped organism. Affected by androgens, male outer privates create from this undifferentiated state or, without any testis, female outside private parts are delivered whether an ovary is available.

4.2. MALE PHENOTYPE

Fetal testicular cells secrete ample testosterone to increase blood concentrations to the same degree as those seen in adult males. Accumulation of testosterone is increased by an additional influence of the gene product TDF gene or SRY (sex-determining region of the Y chromosome), which inhibits aromatase production and prevents the conversion of testosterone to estrogens. Testosterone promotes the growth and differentiation of the Wolffian ducts that develop into the internal male genital tracts.

Sertoli cells in the newly differentiated seminiferous tubule secrete a glycoprotein called antimullerian hormone (AMH) under the influence of the SRY gene product and various transcription factors, inducing apoptosis of tubular epithelial cells and atrophy or reabsorption of the Mullerian ducts (which would have become the female internal genital tract).

The primitive structures that give rise to the outside genitalia in both sexes are the urogenital sinus and genital tubercle. Masculinization of these structures relies on the secretion of testosterone by the fetal testis to form the penis, scrotum, and prostate gland. Those structures grow into the female external genitalia unless stimulated by androgen.

Differentiation is incomplete when there is insufficient androgen in male embryos or too much androgen in female embryos and the external genitals are unclear. Male external genitalia distinction relies on dihydrotestosterone rather than testosterone.

4.3. MALE REPRODUCTIVE SYSTEM

The human male reproductive system constitutes organs that contribute to the reproductive process occured around a male's pelvic region. The key direct function of the male reproductive system is to supply the sperm for fertilization by the male gamete or spermatozoa. The male reproductive system is divided into four main compartments.

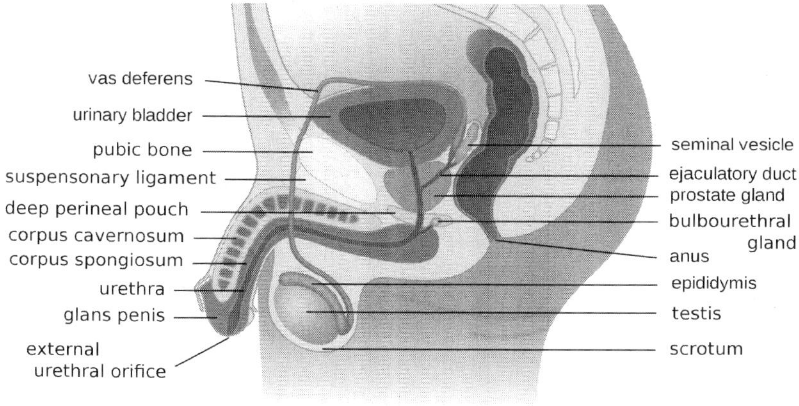

Figure 4.1. Male reproductive system.

Source: Image by Wikimedia

- The testis.
- Accessory ducts: This includes Epididymis, Vas deferens, and Ejaculatory ducts.
- Accessory glands: Accessory glands are internal reproductive organs which supply fluids that nourish the sperm cells and lubricate the duct system. They are the seminal vesicles, the epididymis bulbourethral glands, and the prostate glands (Cowper glands).
- Supporting structures which include the scrotum and the penis.

A male reproductive system includes paired testes, epididymites, ductus deferens, accessory sex glands, and a penis. Spermatogenesis and steroidogenesis are two major roles played by tests. A spermatozoon develops within the seminiferous tubule and testosterone is synthesized within the interstitial compartment of the testis. Spermatogenesis occurs within the seminiferous tubule and steroidogenesis in the interstitial compartment.

Spermatogenesis occurs in stratified epithelial tubules, while testosterone is produced by Leydig cells located between these tubules in an interstitial compartment of loose, vascular connective tissue. The Leydig testis cells, which produce testosterone, play a crucial role in determining secondary sexual characteristics, sperm development, and fertility in men

It is a single, long and extremely convoluted duct that connects the epididymis to the testicular efferent ducts through the vas deferens (a coiled duct). In the transport and storage of testicular spermatozoa, epididymis plays a significant role. Based on their gross morphology, the epididymis is divided into three distinct regions; caput, corpus, and cauda region. The area of the corpus is thinner and it joins the larger segments, caput and cauda.

There is an additional canal in reptiles between the testes and the epididymis head, which receives numerous efferent ducts. However, in both birds and mammals, this is missing. A pseudostratified epithelium surrounds the epididymis. The epithelium is divided from the connective tissue wall, which has smooth muscle cells, by a basement membrane.

In the epithelium, the main cell types are:

Columnar cells, with much of the epithelium in the basal cells. They also have non-motile stereo cilia **Principal cells**, which are long and branching in the head region and shorter in the tail region, extending from the lumen to the basal lamina. Carnitine, sialic acid, glycoproteins, and glycerylphosphorylchlorine are also secreted into the lumen as well.

Basal cells: shorter, pyramid-shaped cells that, before their apical surfaces enter the lumen, touch the basal laminal but taper off. These are known to be undifferentiated primary cell precursors.

Apical cells: These are predominantly located in the head region

Intraepithelial lymphocytes: are distributed throughout the tissues.

Clear cell: Predominant in the tail region. The clear cells in the rat epididymis are subdivided into two types and are concerned with the secretion of either glycoproteins or glycolipoproteins. The blood epididymal barrier, constituted by the zona occludens of the functional complexes at the apical ends of principal cells also appears to play a vital role in maintaining a physiological milieu in the epididymal canal suitable for sperm maturation.

4.4. FUNDAMENTAL COMPONENT OF MALE REPRODUCTIVE ANATOMY

The male reproductive anatomy is divided into five components which are very fundamental to human reproductive health. These include;

4.4.1. Gonadal Development

At eight weeks of gestation (period of pregnancy), Y chromosomes synthesis of H-Y antigen occurs. In the male, this H-Y antigen causes undifferentiated sex glands to develop into testes while in females, lack of H-Y antigen causes undifferentiated sex glands to develop into ovaries.

4.4.2. Duct Development

In this case, both sexes start with two systems such as Mullerian ducts which develop into fallopian tubes, uterus, and inner vagina; Wolffian duct which develops into epididymis, vas deferens and seminal vesicles. In the developmental processes, the male fetal Leydig cells of the testes secrete sufficient testosterone as those seen in adduct men while the Sertoli cells of the testes secrete the anti-mullerian hormone (AMH).

The testosterone (androgen) so secreted is responsible for male sex differentiation during embryogenesis (9th and 13th weeks of pregnancy) and its accumulation is enhanced by an additional effect of the testes determining factor (TDF) gene or sex-determining region of the Y chromosome (SRY) gene product which blocks the expression of aromatase, thus preventing the conversion of testosterone to estrogen.

The testosterone thereby stimulates the growth and differentiation of the Wolffian ducts, which develop into the male internal genital tracts. However, under the influence of the SRY gene product and specific transcription factors, Sertoli cells in newly differentiated seminiferous tubule secrete a glycoprotein called anti-mullerian hormone (AMH), which causes apoptosis of tubular epithelial cells and atrophy or reabsorption of the Mullerian ducts (Which would have become the female internal genital tract).

Notwithstanding, the downstream genes that make up the SRY gene product include the SOX9 and steroidogenesis factor (SF1). These products stimulate the differentiation of Sertoli cells and Leydig's in the testes and also in the formation of tunica albuginea.

4.4.3. External Genital Development

In both sexes, the external genitalia are derived from two primitive structures of the reproductive anatomy. Genital tubercles and urogenital sinuses are included in this. The masculinization of these structures, which produce the penis, scrotum, and prostate gland, requires testosterone released from the fetal testes; in the absence of androgens, these structures develop into the external genitalia of women (clitoris, labia, vaginal opening, etc.).

During embryonic development, when males or females have too much androgen, the external genitalia are ambiguous. Differentiation of the masculine external genitalia depends on dehydrotesterone rather than testosterone. The mechanism through which this occurs is via the conversion of testosterone into dihydrotestosterone (DHT) by an enzyme called 5α-reductase

4.4.4. Brain Development

Sex hormones such as testosterone and estradiol exert their influence during the development of the fetus. Testosterone secreted into the blood reaches the brain and gets converted into estradiol by an enzyme called Aromatase.

The estradiol is what helps in the masculinization of the human brain. In the female, the estradiol secreted by the ovaries binds to a particular protein called α-fetoprotein and therefore prevents its entering into the brain to protect the female brain from being masculinized by estradiol.

4.4.5. Neural Development

A person's brain develops one of the earliest and not the last systems after birth. In utero insult during pregnancy could alter the development of the nervous system due to the development of the most complex structures within the embryo. A neural plate, which folds into a neural groove, forms the early central nervous system.

This early neural is initially open initially at each end forming the neuropores. Failure of these opening to close contributes to a major class of neural abnormalities (neural tube defects). Within the neural tube stem, cells generate the 2 major classes of cells that make up the majority of the nervous system: neurons and glia. Both these classes of cells differentiate into many different types generated with highly specialized functions and shapes.

4.5. ENDOCRINE FUNCTIONS OF THE TESTES

The testes, also known as the male gonads, are male reproductive organs located in the scrotal sacs and are responsible for spermatogenesis. However, various studies have shown that in addition to the spermatogenic functions, the testes also secrete steroid and protein hormones, a role known as the endocrine functions of the testis.

The testis produces androgens such as testosterone (T), dihydrotestosterone and estrogen which are the most typical steroids, and release proteins called inhibins, activins, and anti-Mullerian hormone (AMH)/Mullerian-inhibiting substance (MIS). Collectively these hormones maintain the health of the testes and ensure their proper functioning regarding sperm production and delivery. In the following sections, the androgens and the testicular protein hormones are discussed in detail.

4.6. THE ANDROGENS

Androgens are expected for the turn of events and support major regenerative tissues in men, like the testis, prostate, epididymis, original vesicle, and penis, as well as other male qualities like expanded muscle strength, hair development, etc. To keep an adequate androgen level, androgen development rates should be adjusted against discharge rates and metabolic leeway. The activities of the androgens are affected by the steroid level that can enter target cells, the level of metabolic transformation inside the cells, associations with receptor proteins, and, the activity of androgen receptors at the genomic level. These chemicals control the development of the male regenerative system, as well as the advancement of "manly" actual qualities like whiskers and a profound voice, as well as sexual exercises.

4.7. TESTOSTERONE

A common form of androgen produced by the testes is known as Testosterone which is responsible for the growth of male genitals and sperm production. Testicles in a healthy male can produce about 6 milligrams of testosterone each day and are synthesized and secreted by the Leydig cells of the testis.

These cells do not contain 210-hydroxylase or 110-hydroxylase and further do not synthesize glucocorticoids or mineralocorticoids like the adrenal cortex, which also secrete testosterone. Luteinizing hormone, in conjunction with an adrenocorticotropic hormone produced by the adrenal cortex, increases testosterone synthesis by stimulating cholesterol desmolase, which in turn helps in steroidogenesis.

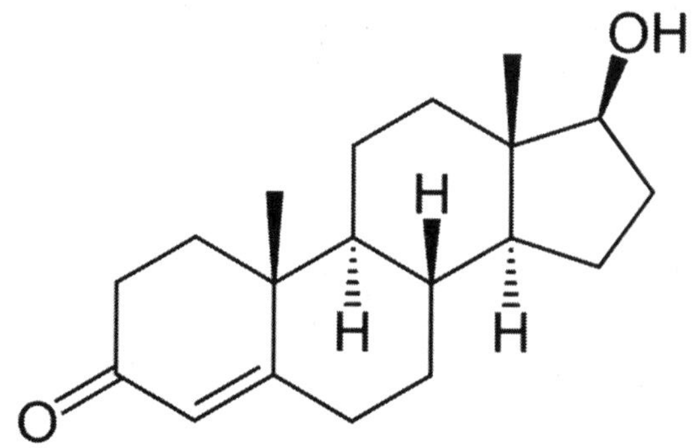

Figure 4.2. Chemical structure of testosterone.

Source: Image by Wikimedia

The prostate contains 5a-reductase which converts testosterone to its active form, dihydrotestosterone. The testicular synthesis of testosterone is controlled by the activities of the hypothalamic-pituitary control mechanism. In this system, the arcuate nuclei of the hypothalamus secrete a gonadotropic releasing hormone (GnRH) into the hypothalamic-hypophysial portal blood which stimulates the anterior pituitary to secrete Follicle-stimulating hormone (FSH) and luteinizing hormone (LH). The luteinizing hormone is then transported to the testis where it stimulates the Leydig cells to produce testosterone.

This action is regulated by the presence of D-aspartic acid that is present in the pituitary gland and the testes and has a role in the regulation and release of LH and testosterone (T). In other words, luteinizing hormone acts on the Leydig cells to promote testosterone secretion, while testosterone acts as an intra-testicular paracrine mechanism to reinforce the spermatogenic effects of FSH on the Sertoli cells as well as the germ cells

4.7.1. Biosynthesis of Testosterone

The androgens are emitted in the testis by interstitial cells of Leydig, which represent 20% of the adult testis mass. Leydig cells are plentiful in both infant and adult males. Notwithstanding, in youth, these cells are scant or non-existent. Subsequently, androgen secretion happens in babies and after adolescence. Its secretion starts in the seventh seven-day stretch of fetal life by the fetal genital ridge.

Around the second to fourth month of fetal life, the testicles start to discharge testosterone. Human chorionic gonadotropins, which are emitted by the placenta during pregnancy, stimulate testosterone discharge from the testicles. Be that as it may, until the age of 10 to 12 years of age, basically no testosterone is discharged.

Following that, testosterone discharge starts, and it quickly increments at the beginning of adolescence and goes on until the end of one's life. Following 40 years, the secretion starts to decline and arrives at almost no by the age of 90.

Cholesterol, a substrate that can be combined from acetic acid derivation once more or taken up from plasma lipoproteins, is a forerunner in the combination of steroids at any stage. The low thickness lipoprotein part has all the earmarks of being the essential extracellular store of cholesterol in human Leydig cells.

Besides, intracellular lipid beads containing cholesterol esters can act as intracellular cholesterol stores. The Cytochrome P450 side-chain cleavage protein of the inner mitochondrial membrane of Leydig cells hydroxylates the side chains C22 and C20 of cholesterol to change it into pregnenolone. It is then translocated to smooth ER for conversion to testosterone in two pathways.

4.7.2. Actions of Testosterone

The hormone helps in the differentiation of the epididymis, vas deferens, and seminal vesicle. It is also responsible for the pubertal growth spurt and the cessation of the pubertal growth spurt (epiphyseal closure). It ensures libido, spermatogenesis in Sertoli cells (paracrine effects), deepening of the voice in males, increased muscular mass, growth of the penis and seminal vesicles, and negative feedback control of the anterior pituitary. To carry out its role in spermatogenesis, testosterone is said to mediate the maintenance of the blood-testis barrier (BTB).

4.7.3. Degradation of Testosterone

Many target tissues convert testosterone into dihydrotestosterone, the most active androgen. Some tissues, including adipose tissue, the hypothalamus, and the liver, convert testosterone to oestradiol. The liver degrades the majority of testosterone into inactive androsterone and dehydroepiandrosterone that are then conjugated and excreted in the urine

4.8. PHYSIOLOGIC ROLES OF AMH IN MALES THROUGHOUT LIFE

Along with inhibin B and FSH, the Mullerian chemical is supposed to be the earliest chemical discharged by the Sertoli cells in males. Subsequently, the chemical is a significant mark of Sertoli cell capacity and it assumes a crucial part in the turn of events and elements of the male reproductive system.

For instance, this chemical is fundamental for fetal sex separation all through the pre-birth period. Before the seventh seven day stretch of pregnancy, male and female embryos have separate portions, bipotential outside genitalia, and two sets of unipotently channels (the Müllerian ducts and the Wolffian ducts). In the XY baby, the expression of the SRY gene, the sex inversion quality on the Y chromosome, causes testicular separation.

From the eighth seven day stretch of pregnancy onwards, the fostering testicles' physical cells, Leydig cells, and Sertoli cells secrete testosterone, Insulin-like factor and AMH are essential for the singular's typical male separation. Even though testosterone has been examined before, it is critical to state here that it advances the separation of the Wolffian ducts into original vesicles, vasa deferential, and the epididymis during fetal turn the of events.

Insulin-like factor 3, additionally produced by Leydig cells, goes about as a significant sign during the main period of testicular plunge. The relapse of the Mullerian ducts prompting the separation of the genitalia into separated into the oviducts, uterus, and upper part of the vaginal canal in the female embryo is because of the activity of AMH.

This GnRH flood causes a huge expansion in gonadotropins (LH and FSH) in the initial not many long stretches of life, trailed by an expansion in degrees of testosterone, AMH and inhibin B. AMH levels stay high all through the prepubertal time of life and are decreased during pubescence as testosterone levels rise.

During prepubertal periods, Leydig cells produce low amount of testosterone and Sertoli cells are yet not fully developed, making the course of development of spermatozoa captured in a premeiotic stage while AMH remains secreted at a normal amount till the beginning of pubescence. Leydig cells in men go through extra separation toward the beginning of adolescence following the secretion of GnRH and the resulting "LH flood".

There is additionally an expansion in testosterone biosynthesis, which prompts the development of Sertoli cells. Sertoli cell development makes germ cells go through meiosis, which begins the course of sperm synthesis.

The inhibitory activity of testosterone wins over FSH stimulation, bringing about a reduction in AMH expression and the resulting decline in its circling levels. In adult males, AMH secretion arrives at its peak and remains almost consistent until the end of a man's life.

4.9. SEMINIFEROUS TUBULES AND SPERMATOGENESIS

Testes or testicles appear as a pair of oval-shaped organs enclosed in the scrotum and situated behind the penis and in front of the anus. They produce male reproductive cells, spermatozoa, and androgens, the male hormones. Each adult testis weighs 12 to 19 g, 4.5x 2.5x 3 cm in dimension and is suspended in the scrotum by a spermatic cord. The rete testis at the mediastinum of the testis connects to the head of the epididymis, which is opposed to the testis posteriorly.

The tunica albuginea, fibrous capsule, covers each testis. The testis is divided by partitions of the tissue from the tunica albuginea into approximately 250 lobes. Three to ten coiled tubules are inside each lobe. These tubules are called seminiferous tubules containing two different cells population: spermatogenic or germ cells and Sertoli cells surrounded by a peritubular myoid cell layer. The stroma between the seminiferous tubules is called the interstitial (interstitial tissue), where blood and lymphatic vessels, the steroidogenic Leydig cells and other cell types are located.

4.9.1. Tunica Albuginea

Each testis is enclosed in a thick fibrous envelope, formed by collagen fibers impregnated with elastic fibres (5% elastin), called tunica albuginea. Tunica albuginea is formed by two layers: the outer longitudinal layer and the inner circular layer.

Because of its contractile properties (erection), the tunica albuginea has different physiological functions: the preservation of the interstitial pressure inside the testis, the support of the spermatozoa movement from the testis to the epididymis, and the regulation of the blood movement through the testis. On the posterior surface of the testis, the tunica albuginea becomes thicker to form the mediastinum testis from which Septula testis enter the gland, separating it into almost 250 testicular lobules.

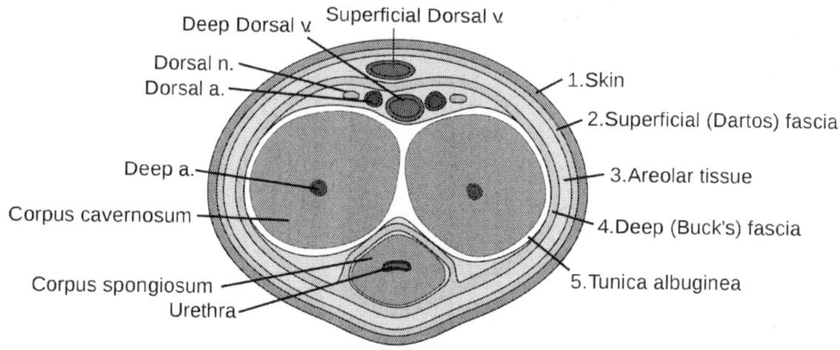

Figure 4.3. Cross-section image of Penis.

Source: Image by Wikimedia

4.9.2. Basement Membrane

The basement membrane is a fibrous matrix formed by type IV collagen, glycoproteins and Lamin produced by the epithelial cells. It plays a crucial part in keeping up the structural and functional integrity of tissues in the testis.

Modified cellular layer structure has been related to extreme function abnormalities of the testis like cryptorchidism, autoimmune orchitis, and vasectomy.

4.9.3. Peritubular Cells

In human testicular, the outer coat of the seminiferous tubules is formed by several layers of myoid, peritubular cells and extracellular matrix (ECM) proteins. The cells are peritubular myofibroblast-like cells that encompass the seminiferous tubules to maintain their structural integrity and are capable of tubular contractility and sperm transport.

These cells, in adults, express markers for smooth-muscle-like cells similar to the smooth muscle actin. Immunohistochemical studies suggested that the cellular phenotypes differ between the outer and inner layers. After stain, the inner layers showed a smooth muscle phenotype stain with desmin. While the outer layers were stained with vimentin indicating a connective tissue phenotype.

A basal lamina separates the spermatogonia stem cells (SSC) and the peritubular myoid cells (PMC). This can indicate a possible cellular interaction between the PMCs and SSC to maintain the SSCs niche, similarly to Sertoli cells. One of the contribution mechanisms is through the production of secreted factors like glial cell line-derived neurotrophic factor (GDNF), which acts in combination with the androgen receptor (AR).

It has also been demonstrated that the PMCs produce colony-stimulating factor 1 (CSF-1) in interaction with specific receptor CSF-1R to regulate the SSCs activity.

4.9.4. Leydig Cells

Leydig cells are called also interstitial cells because they are in the stroma between the seminiferous tubules: the interstitial (interstitial tissue) holding the tubules together within each lobule. This tissue is activated at puberty through the interstitial-cell-stimulating hormone of the anterior lobe of the pituitary gland.

Aftcr stimulation through the luteinizing hormone (LH), the production of testosterone via the Leydig cells increases through the regulation of the expression level of steroidogenic enzymes like the 17-β hydroxysteroid dehydrogenase.

Testosterone exerts its effects locally by binding to the androgen receptor (AR) within the testis or distantly by binding to androgen binding protein (ABP) which increases its levels in the seminiferous tubules and carrying to the epididymis.

Elevated levels of serum LH, as well as FSH and lowered levels of serum testosterone suggest Leydig and germ cell failure.

4.10. STRUCTURE AND FUNCTION OF THE SEMINIFEROUS TUBULES

The seminiferous tubules are the fundamental units of the testes where the SSCs multiply and separate through cell cycles (mitosis, meiosis, post-meiotic spermatid development, and spermiogenesis) to create spermatozoa in an interaction called spermatogenesis.

In people, the seminiferous tubules address around 60% of the total gonad volume and they are around 200 μm in width and have a complete length of ~600 meters. These seminiferous tubules are made from the lamina propria (peritubular tissue) with around 80 μm level and the germinal epithelium with around 8 μm thickness.

The germinal epithelium is made from enormous Sertoli cells and spermatogonia germ cells. These cells are associated using tight junctions.

The phase of the seminiferous epithelium cycle has an impact on the structure of seminiferous tubule areas. Likewise, the nerves, lymph vessels and veins don't enter the seminiferous tubule and are found exclusively on interstitial tissue.

The seminiferous tubules have terminal closures in the mediastinum testis and are empty through straight cylindrical expansions called Tubule seminiferous rect.

4.10.1. Sertoli Cells

Enrico Sertoli was the author of the first publication reporting the existence of Sertoli cells. Later, numerous reviews in scientific journals and books have been published describing the Sertoli cell morphology and functions, mostly focusing on mammals.

In humans, Sertoli cells are crucial for testis physiology. They proliferate during the perinatal and neonatal period, becoming quiescent for several years and having a second peak of proliferation just before puberty.

Although, around puberty Sertoli cells stop proliferating and start to differentiate, being therefore able to support full spermatogenesis. The establishment of the Sertoli cell barrier and fluid secretion/lumen formation is the clear character of Sertoli cell maturation.

Follicle-stimulating hormone (FSH) and androgens are considered important factors that regulate Sertoli cell proliferation. In addition, estrogens, activins, TGF-beta, BMPs, interleukins and TNF- alpha are factors involved in the proliferation and differentiation of Sertoli cells.

Sertoli cells were identified as 'nurse cells' because they are morphologically reshaped by the developing germ cells and have a multitude of cytoplasmatic processes. Each Sertoli cell is "nursing" approximately 30–50 germ cells at four or five diverse stages of their advancement at any given time throughout the epithelial cycle.

Structural characteristics of the Sertoli cells vary among species, such as the heavily vacuolated nucleolus present in some ruminants, the nucleus localization in the middle of the seminiferous epithelium in monkeys, the presence of Charcot-Botcher crystalloids in men, and the presence and amount of lipid droplets and glycogen in the Sertoli cell cytoplasm.

Therefore, the Sertoli cell shape may vary according to the species and the progression of spermatogenesis and the tasks. As the germ cell requirements changed, interactions and metabolic needs change substantially and accordingly, high variations are detected in the Sertoli cell cytoplasm extension, the number of nuclear pores, the presence and translocation of organelles and the protein expression pattern and location across the different phases of spermatogenesis.

On the other hand, the Sertoli cells are considered "epithelial" cells as they are based on a strikingly thick basal lamina, appear a remarkable design (polar-basolateral-apical) with horizontal cell-cell junctions and border on a luminal space.

Although Sertoli cells extend from the basement membrane of the seminiferous tubule into the abluminal compartment, the two tubular compartments are isolated by tight and adherent junction complexes between neighboring Sertoli cells, which work as the major component of the blood-testis barrier (BTB). These junctions generate the required chemical environment for the fulfillment of meiosis and spermiogenesis.

Besides, the molecular character of Sertoli cells changed from keratin IFs to vimentin IFs during their development and maturation. Also, a wealth of special and rather extended forms of adh junctions connected Sertoli cells and spermatogenic cells instead of the typical epithelial junctions.

Functionally, they play a critical role during spermatozoa rent development by supporting and organizing spermatogonia germ cells during different stages of spermatogenesis through secretion of androgen-binding protein and interaction with Leydig cells. In addition, they provide the germ cells with a variety of ions, nutrients, carbohydrates, hormones, and growth factors.

4.10.2. The Transition Region

The seminiferous tubules connect to the rete testis in a region named: the Transition region. This region might be a specific area for immature Sertoli cells. Likewise, the transition region contains a subpopulation of mitotically dynamic Sertoli cells without separation, Sertoli cells markers like record factor GATA-4 and the androgen receptor. It tends to be expected that the adult Sertoli cells populace isn't morphologically homogeneous.

As the transition region presents adjusted Sertoli cells that show includes that look like undifferentiated Sertoli cells, with fewer spaces, more modest nucleolus, and more heterochromatin. Consequently, the doctrine that the adult Sertoli cells populace is terminally separated in well-evolved organisms has been tested by a few late examinations.

Moreover, because this transition region of mammalian testis additionally contains spermatogonia undeveloped cells, it has been guessed that it may be a region where the seminiferous tubules keep on filling in physically mature people. Others showed that the transition region is a site where seminiferous tubules are initially formed.

4.11. POSITIONAL RELATIONSHIPS AMONG MALE REPRODUCTIVE ORGANS IN INSECTS

The morphology, structure and size of organs have useful importance. Insects are the most plentiful of all organisms as far as species number, bringing about female and male reproductive organs being exceptionally different in their structure. While the general example of spermatogenesis in insects is fundamentally like that in well-evolved organisms, the morphology, design, and size of sperms in insects are profound.

Since available sources are generally restricted, the quantity of sperm delivered ought to be conversely corresponding to their size. Albeit more modest testicles don't be guaranteed to deliver numerous sperm, sperm size is firmly connected with testis size.

Among Drosophila groups, there are positive connections between testis length and sperm length. Besides, sperm length connects emphatically with body size in butterfly species. How much sperm a female gets and stores in the spermatheca (female sperm storage organ) ought not to be entirely set in stone by the number of sperm utilized for treatment during the post-mating time frame. Numerous insect species are known to over-discharge under research facility conditions yet it is muddled whether the unreasonable discharge is normal in the field.

As a rule, sperm size in organisms including insects isn't relative to body size. Drosophila flies, as is notable, are minuscule and consequently have few testicles, albeit *D. bifurcate* has enormous testicles containing extremely lengthy sperm arriving at a length of 5.8 cm. Accordingly, the quantity of huge sperm that they produce is little.

This might be because of the size of the testicles as well as other reproductive organs. Monster sperm are looped, yet their length and thickness should not impede the development inside the male and female reproductive system. It is felt that the expense of sperm creation, as well as the control of sperm relocation, would be limited to make the gamete bigger, however as a matter of fact some Drosophila sperm have become huge past this imperative.

Sperm delivered in the testicles are generally put away in the fundamental vesicles, assuming they are available. In insects, when present inside the male regenerative organs, sperm are either not motile or their motility is more suppressive than when they are held in the female reproductive organs. Accordingly, guys need to both store sperm in the fundamental vesicles, and breaking point the energy costs related to those sperm until mating.

Late examinations have uncovered that sperm age or pass on in the spermatheca or male regenerative organs, and it is fascinating to perceive how maturing influences sperm quality and endurance in the fundamental vesicles because the morphology of the fundamental vesicles might be related to sperm maturing, as referenced underneath.

Furthermore, a recent molecular biology study indicates that increased expression of seminal fluid protein incorporated in spermatophore genes is correlated with increased sperm viability in the ejaculates.

4.12. FUNCTION AND STRUCTURE OF TESTES

Testicles work to deliver sperm and thus consume or discharge the different substances including supplements and chemicals for spermatogenesis. In species lacking original vesicles, mature sperm are put away in the testis close to the vas deferens. The state of the testis is round, oval, or prolonged, maybe relating to sperm length, presumably because sperm grow significantly during spermiogenesis (from the spermatid to the sperm).

The testicles are generally matched and the numbers change broadly from species to species. The shade of the testicles likewise changes enormously relying upon the species; for instance, numerous lepidopteran insects have white, yellow, red, or purple testicles. Albeit the colors that are not generally required in the body might be kept in their testicles, their capacity and transformative importance have not been explained.

In *P. c-aureum*, the testis is covered with a yellow membrane from the last instar larval stage to the early adult stage while spermiogenesis is actively happening. Albeit the synthesis and capacity of this membrane are obscure, it is conceivable that it effectively safeguards the testis and sperm cells from bright light or potentially different substances in the hemolymph, that are unacceptable or harmful to spermatogenesis.

Alternatively, this membrane may positively improve the nutritional and/or humoral conditions for spermatogenesis. Interestingly, the time when this membrane begins to degenerate coincides with the time when adult development of wings is almost complete, the scale of spermatogenesis and testis size begin to shrink, and sperm begin to migrate from the testes to the vas deferens.

These simultaneous occasions are emphatically thought to be related to chemicals in the hemolymph, for example, ecdysteroid. Curiously, it is displayed in Calpodes ethics Stoll (Lepidoptera, Hesperiidae) that the testis is encircled by a yellow shade

Testis development including spermatogenesis is impacted by formative stages, temperatures, dietary circumstances, and chemicals. In the yellow manure fly *Scathophaga stercoraria* (L.), the testicles decay in the wake of mating. Likewise, in *Drosophila melanogaster* Meigen, there was a huge decrease in the size of the testicles after five progressive matings.

As a rule, testis development is firmly connected with spermatogenesis, and testis size normally increments as spermatogenesis turns out to be more dynamic. In any case, the connection between testis size and spermatogenesis relies upon the species and formative stage.

4.13. SPERM POLYPHENISM

Sperm polymorphism is apparent in the various insect orders. Generally, one sperm is long whereas the other is short; giant sperm have been observed in some Drosophila and beetles. Giant sperm fertilize the eggs and short sperm do not participate in fertilization, although contradictory findings raise the possibility that both types of sperm can fertilize.

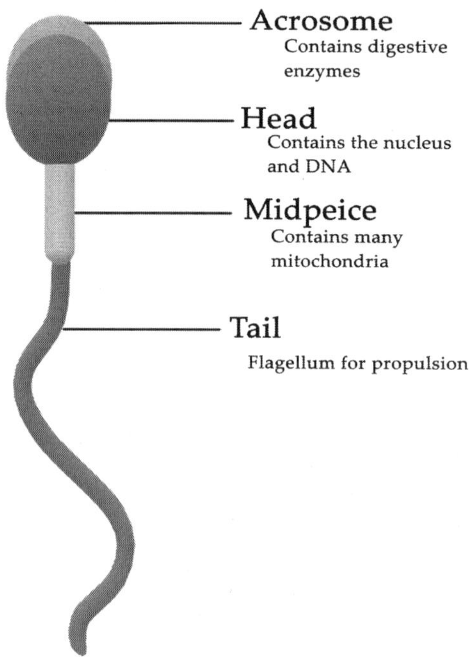

Figure 4.4. Sperm anatomy.

Source: Image by Wikimedia

In lepidopteran insects, there are fertile nucleated eupyrene sperm and infertile non-nucleated apyrene sperm. Eupyrene sperm fertilize the eggs, whereas apyrene sperm cannot fertilize the eggs because they lose their nuclei during meiosis. In general, apyrene sperm are produced more or transferred to females than are eupyrene sperm, although spermatogenesis of apyrene and eupyrene sperms is not markedly different in the diamondback moth *Plutella xylostella*. This seems to be related to the fact that apyrene sperm are shorter than eupyrene sperm and the former is generated later in development.

4.14. STORAGE AND MIGRATION OF SPERM

The production and storage of sperm can be a lifelong event for males. There are four types of sperm migration within the male reproductive organs. First, sperm formed in the testis move to the vas deferens or the seminal vesicle. In lepidopteran insects, sperm migration occurs from the testis to the duplex via the vas deferens, with a circadian rhythm. Second, sperm move from the seminal vesicle through the ejaculatory duct to the female reproductive tract during mating.

Third, there is a process called sperm reflux. In *P. c-aureum*, the accessory glands open into the duplex, but not into the ejaculatory duct, thus when the accessory gland material passes the duplex, it ejaculates all sperm present. This prevents sperm from being preserved in the next copulation. Thus, this butterfly regurgitates excess sperm in the duplex into the vas deferens during mating.

Finally, sperm migration sometimes occurs immediately after mating. Male sweet potato weevils *Cylas formicarius* (F.) can mate several times a night, and sperm migrate from the testes to the seminal vesicles immediately after mating. This may be because the seminal vesicles are adjacent to the testes, which spermatogenesis is active during the adult stage, and that sperm production is completed in the testes. Thus, the positional relationship between the testes and the seminal vesicles, the degree of sperm perfection in the testes, and the stage at which sperm are formed determine the male ejaculation pattern.

4.14.1. Reproductive Toxicology: An Update

Reproduction (procreation; conception) is one of the most essential requirements of all organisms where producing a transcript, helps in the survival and perpetuation of the species. Parenthood is one of the most comprehensively preferred priorities of mankind and happens at ease when couples are vigorous and normal. According to earlier archaic studies by anthropologists and evolutionary biologists, *Homo sapiens* displayed a better cognitive development about 2,00,000 years ago and hence, had a "reproductive consciousness".

Ancient mythologies and civilizations respected and worshipped fertility Gods and Goddesses like the Egyptian goddess Maat, Mesopotamian's Erua, Babylonian's Ishtar, Persian's Anaitis, and Greek's Actmia, and also, they had a deep desire for conception, and a strong perception of fertility, which can be correlated with human sustainability and existence.

The reproductive cycle of a mammalian individual involves a sequence of several phases and unified events. According to ICH guidelines:

- Pre-mating to birth (mature male and female reproductive ability, growth and maturation of gametes, reproductive nature, and conception).
- Conception to implantation (mature female reproductive ability, pre-implantation development, cleavage, morula, blastula, and implantation).
- Implantation and organogenesis (mature female reproductive ability, development of embryo, and foremost organ development).
- Fetal development (until the end of gestation, mature female reproductive ability, fetal development and growth, and growth and development of organs).
- Birth and pre-weaning development (mature female reproductive ability, parturition, lactation, neonate adaptation to extrauterine life, pre-weaning development, and growth).
- Post-weaning development up to sexual maturity (growth, adulthood, adaptation to independent life, and achievement of full sexual function).

According to World Health Organization (WHO) info, at present globally 50–80 million people are facing infertility. Significant studies have reported that female infertility occurs in 50%, infertility because of male factors is 20–30%, and the rest is shared by both genders. These findings are considerably broader than previously reported.

4.14.2. Reproductive Toxicity

For a decade, human regenerative interruption by different elements including xenobiotics like medications, work-related, and environmental exposures prompting reproductive damage which has turned into a developing concern. Reproductive damage is characterized as: "the opposing impacts of a substance on any qualities of the male or female sexual regenerative cycle, along with a disability of reproductive capacity, and the enlistment of unfavorable impacts in the undeveloped organism, like development issues, deformities, and mortality which would disrupt the creation and development of typical posterity that could be raised to sexual development, able thusly of imitating the species".

The primary fundamental presentation in reproductive toxicology was given by Wilson and Warkany in 1965. The principal test rule was distributed by the Food and Drug Administration (FDA) in 1966, followed by the Committee on Safety of Medicines, the Ministry of Health and Welfare (MHW) of Japan, and the rest of the different countries.

It was temporarily ended by the International Federation of Teratology Societies (IFTS), the drug industry and the health specialists of EEC, Japan, and the USA with the guide of ICH Harmonized Tripartite Guideline "Identification of Toxicity to Reproduction for Medicinal Products" in June 1993.

4.15. TESTICULAR HISTOPATHOLOGY AND SPERMATOGENESIS IN MICE WITH SCROTAL HEAT STRESS

High environmental temperatures influence the functioning of numerous natural systems in the human body, like the circulatory system, integumentary system, and respiratory system, with results in male reproductive action.

The testicles' temperatures of most various types of well-evolved organisms are in many cases lower than the internal heat level for typical spermatogenesis. An increase in testicular and epididymal temperatures in men and different vertebrates prompts diminished sperm yield, and diminished sperm motility, and builds the level of sperm with abnormal morphology.

Increases in testicular temperature may be detrimental to spermatogenesis and ultimately cause problems infertility. Elevation of scrotal temperature interrupts spermatogenesis, resulting in reduced the number and motility of spermatozoa, fertilization ability of the surviving sperm, and poor fertilization-embryo.

A high-temperature surrounding, even though it could be inside as far as possible, harms sperm quality. The ideal testicular temperature for spermatogenesis is from 2 to 4°C lower than the internal heat level. Each 1°C expansion in testicular temperature prompts a 14% reduction in spermatogenesis. As of late, research by Gong et al. showed that high encompassing temperature lessens sperm motility through diminished mitochondrial movement and ATP formation.

Heat-stressed testicular tissue prompts apoptosis through mitochondrial pathways or DNA damage. Plus, damaged sperm in the vas deferensaffected by high surrounding temperature can induce male sterility. Consequently, heat-stress is a high-risk factor influencing testicular structure and diminishing sperm quality.

This causes oxidation, bringing about damaged thermoregulation of the gonad, for example, a decrease in of antioxidant compounds and creation of heat shock proteins, expansion in apoptosis and cell death. Heat decreases the capacity to manage the temperature of the scrotum, causes oxidative stress, in this manner causing cell responses including mitochondrial damage, increases receptive oxygen species (ROS) levels, and diminishes the formation of ROS scavengers.

Scrotal heat stress causes an assortment of components that happen in the testicles including oxidative damage reaction, heat shock reaction, cell cycle designated spots, DNA damage, apoptosis, and cell death. Scrotal heat stress hinders spermatogenesis in male mice. Spermatogonial germ cells are likewise affected by heat stress, bringing about the destruction of them in the seminiferous tubules, Sertoli and Leydig cells are declined.

Spermatogenesis is related to many elements, including temperatures that can cause cell and sub-molecular changes, influencing gene expression that upsets sperm formation, bringing about diminished reproductive wellbeing. Germinal cells are dynamically sensitive to high temperatures Critical apoptotic loss of germ cells after testicular heat stress might happen either through internal or external pathways.

Cells in the testis of vertebrates are impacted by heat stress essentially spermatocytes and early spermatids. The testicles of the mice in the heat stress group showed degenerative changes with spermatic capture in a large portion of the seminiferous tubules. DNA damage in sperm was seen in the testicular tissue presented to heat pressure.

The biomechanism of sperm DNA damage involves various types of oxidative reactions, DNA repair errors in the late stages of spermatogenesis, and functional abnormalities that reduce the protective ability of Sertoli cells leading to increased DNA fragmented sperms.

Many factors contribute to increasing testicular temperature. These factors can be grouped according to habits, lifestyles, occupational factors that must be exposed to high temperatures for long periods, and climate change. Depending on the situation, the testicles may experience transient or prolonged heat stress with varying intensity. The effects of high ambient temperature on male fertility tend to accumulate with repeated exposure over a while.

Mice are animal models ordinarily utilized in biomedical exploration, since they are not difficult to deal with, financially affordable, and have similitudes to indoor regulators like a humans. A few investigations have utilized such mouse models to assess the effect of heat on spermatogenesis.

Past exploration has demonstrated the way that germ cell death can be actuated considering scrotal temperature in mice. Nonetheless, the animal model for heat stress study isn't yet deeply grounded.

The aim of this part is to depict a mouse model for researching the testicular histomorphometric change because of testicular intensity stress and examine the adverse consequence of persistent scrotal heat pressure on the human male reproductive system.

4.16. METHOD TO GENERATE A MOUSE MODEL FOR TESTICULAR HEAT STRESS

4.16.1. A Mouse Model for Testicular Heat Stress

Swiss mature male mice 8–10 weeks old (20-23 g) were kept in an animal facility at a controlled temperature (25 ± 1°c) and illuminated (12 h light-12 h dark), and with free access to food and water.

The mouse model was established to study the effect of scrotal heat stress on spermatogenesis and male fertility. The lower body (including the scrotum) of the mouse was soaked in a thermally controlled water bath (heat exposure at 37°C, 40°C, or 43°C) 2 times a day 10 minutes apart, every 10 minutes, lasting for 5 consecutive weeks, 6 days a week. Control mice were treated in the same way, but in a water bath maintained at room temperature.

After having the bath, mice were dried and examined for any injury or redness to the scrotal skin before being returned to their cages. Mice were kept in different cages and had free access to water and food. All animals are cared for under identical environmental conditions and monitored in their general health.

4.16.2. Tissue Processing and Hematoxylin-Eosin Staining

After completing the heat exposure experiment after 5 weeks, all mice were sacrificed under anesthesia. Their testis tissue was harvested for histological morphometric analysis. The testicular specimens were individually immersed into 4% buffered formaldehyde and dehydrated with graded concentrations of ethyl alcohols at room temperature.

The testicular specimens were then embedded into paraffin. The paraffin blocks were cut thinly with a thickness of 5 μm and transferred into gelatinized slides. The sections were deparaffinized with xylene and then rehydrated through a descending series of ethanol and water. Slides were stained with hematoxylin and eosin (H&E) and observed under a light microscope for histopathological analysis.

4.16.3. Johnsen's Mean Testicular Biopsy Score Count

Testicular histological damage and spermatogenesis were assessed using Johnsen's mean testicular biopsy score under light microscopy. Thirty tubules for each testis were graded and each tubule was given a score from 1 to 10 based on the presence or absence of germ cell types in the testicular seminiferous tubules such as spermatozoa, spermatids, spermatocyte, spermatogonia, germ cells, and Sertoli cells to evaluate histology.

A higher Johnsen's score indicates a better status of spermatogenesis, while a lower score refers to more severe dysfunction. A score of 1 means no epithelial maturation is considered for the tubules with complete inactivity while a score of 10 means full epithelial maturation is considered for the tubules with maximum activity.

4.17. THE EFFECT OF HEAT STRESS ON MALE REPRODUCTION

Genital heat stress has been known as a gamble factor for male infertility. In any case, the specific component causing weakened spermatogenesis is as yet muddled. Testicular histology after warm openness had changed like diminished epithelial thickness, the presence of cell flotsam and jetsam, divided cells, and nonappearance of sperm and spermatocytes.

The effect of drawn-out heat pressure adversely influences sperm quality and amount, testicular structure with disturbance of the encompassing epithelium. What's more, scrotal intensity stress lessens testicular weight, mitochondrial degeneration, dilatation of the smooth endoplasmic reticulum and Leydig cells lose the capacity of supporting immature germ cells.

Discoveries from this review support that the testicular design confused germinal epithelium with checked sloughing or decimation of the lumen; spermatogenic interfered red with the shortfall of many sorts of sperm cells in male mice after ongoing scrotal intensity stress uncovered for a long time.

The raised testicular temperature in vertebrates impedes spermatogenesis, germ cells cause harm, diminish sperm motility, and expand the level of sperm with an unusual morphology. Paul et al. likewise showed the impact of heat on testicular capacity and diminished fertility in mice. In addition, heat stress increases free radicals and oxidative stress resulting in apoptosis of germ cells and increased sperm DNA fracture, loss of sperm integrity.

Many elements can increment testicular temperature. These variables that influence male ripeness can be assembled by propensities, ways of life, word related factors that should be presented to high temperatures for significant stretches, and environmental change. Contingent upon the circumstance, the gonads might encounter transient or delayed heat pressure. The impacts of high temperature on male fertility will require study over some time.

Heat-stress has a direct impact on work performance by increasing the risk of illness and work-related injuries. When workers are exposed to high temperatures, their bodies are unable to activate compensation regimes or recover from stressful working days that put their health in danger.

A researcher named Boni demonstrated high ambient temperature as a serious threat to reproductive function in animals and humans. Comparing parameters between the control group and the group workers exposed to heat were within the limits of normozoo sperm range, there was a serious decline in semen parameters. The bakers expose to high ambient temperature have high infertility rates as shown in Al-Otaibi's study.

There was another large-scale epidemiological study at the Danish infertility clinic in which subjects underwent sperm examinations or infertility treatments, and obtained information on occupational exposure. It was found that the groups of workers exposed to textile dyes and lead, noise, cadmium, and mercury were all potentially infertile.

Similarly, Hamerezaee et al. have drawn the same results about the effects of temperature stress on semen quality when studied on workers of the steel industry in Iran. The semen quality of the workers is lower than the normal limit and significantly lower than the control group.

Another three-year study conducted in a fertility clinic in New Orleans found that men working in buildings without air conditioning during summertime would reduce their sperm quality. Their semen samples had a significantly lower sperm concentration, number of sperms per ejaculation, and a lower percentage of sperm motility when compared those figures in summer to other seasons of the year. In contrast, a study in the ceramic industry has reported that workers exposed to high temperatures had no significant difference on semen analyses except for the sperm velocity.

Taking everything into account, this section shows the proof of adverse consequences on histopathological modifications and spermatogenesis following persistent scrotal heat stress. This part additionally describes a model for concentrating on the male reproductivesystem and normalized Johnsen scores system to survey murine testicular histopathology in the seminiferous tubules.

4.18. METHODS OF SPERM SELECTION FOR *IN-VITRO* FERTILIZATION

Male infertility accounts for 50–60% of infertility cases and abnormal semen qualities like low motility, low sperm concentration, abnormal morphology and increased levels of sperm DNA damage are characteristic of infertile men samples. Furthermore, a high level of reactive oxygen species (ROS) is found in 40–88% of sperm samples of infertile men and physiological sperm functions such as capacitation, acrosome reaction and hyperactivation require low ROS concentration while ROS overproduction is usually due to the inability of antioxidant to neutralize ROS.

Also, decrease sperm motility, DNA integrity and viability, and increase midpiece defects are caused by oxidative stress from a high level of ROS and decreased levels of antioxidants. In addition, lower *in vitro* fertilization pregnancy rate, irregular preimplantation development, early loss of pregnancy and decreased rate in ART conceived offspring are correlated with poor DNA integrity.

A result of the above heterogeneity of ejaculate (understanding of sperm physiology) and male gamete integrity rule in both fertilization and embryogenesis, has led to increased demand for sperm selection techniques. Sperm biology, sperm concentration, volume and lifetime *in vitro* are the fundamental challenge of sperm selection and the sperm selection process's ideal time is about 10 minutes for 1 ml of sperm sample containing 100 million/ml. This shows a very high biological sorting rate of −100 KHZ and the current cell sorting technologies has a lower value than this.

There are various sperm selection techniques in Assisted Reproductive Technology (ART) and these techniques try to replicate the natural process (*in vivo*) in which quality sperms are selected from other constituents of the ejaculate as they actively move through the cervical mucus.

4.18.1. Sperm Production

At discharge, semen comprises a suspension of spermatozoa which is put away in the epididymis and is blended in with the secretion of the embellishment organs. The prostate and the fundamental vesicles are the principal organs while the bulbourethral organs and the epididymis are answerable for the minor commitment of the discharge.

The original liquid comprises a prostatic portion that is rich in sperm cells and the vesicular division which is less in spermatozoa. Besides, it is fundamental to have a total example volume assortment and not to lose the main rich sperm portion during masturbation which could make semen examination troublesome. In this manner, complete semen test creation is the initial step all through sperm planning.

In addition, sample production is carried out through masturbation with sterile specimen container. This is done after abstinences of 2–3 days which maximize conception. Furthermore, patient should be encourage onsite production of semen sample which avoids extreme temperature exposure while offsite semen production should avoid spermicidal effect from lubricant and samples should get to the andrology/*in vitro* fertilization (IVF) laboratory within 30–40 minutes without extreme environmental temperature exposure.

4.18.2. Choice of Sperm Selection Techniques

The selection or preparation techniques use in sperm separation in an IVF laboratory depends on the characteristic of the semen sample when subjected to semen analysis. Furthermore, sperm cells should be separated from seminal plasma as early as possible and *in vitro* fertilization capacity diminishes permanently when sperm cells are not separated from seminal plasma within 30 minutes of ejaculation.

Furthermore, World Health Organization recommends sperm cell separation from seminal plasma within one hour of ejaculation and this will limit damage from leucocytes and other cells present in the semen. Semen samples with severe oligo and asthenozoospermia are separated using the simple wash and normozoospermic samples are separated with swim-up or density gradient. Also, suboptimal quality semen samples are separated by density gradient

4.18.3. Simple Wash Method

It involves one or two centrifugation rounds of semen samples to separate the sperm pellet from the seminal plasma. This process does not significantly decrease sperm count, normal morphology, and motility remain unchanged but there is an increase in rapid forward progression and hypermotility of sperm cells in the post-washed sample.

Furthermore, this process is used for cases of severe oligospermia, and asthenozoospermic semen samples and it is recommended method for insemination-ICSI and not for standard insemination procedure in IVF. Also, the centrifugation process in this method causes additional harm to sperm cells by the production of reactive oxygen species (ROS) by leucocytes and abnormal sperm cells. ROS production causes DNA damage in spermatozoa, decreased sperm motility, increase the number of apoptotic sperm cells and decrease sperm plasma membrane integrity.

Method:

- The semen sample is mixed well.
- Add supplement media to semen sample in a ratio of 1:1 in a 15 ml conical tube.
- Centrifugation is done at 1800 g for 5–10 minutes.
- Aspirate the supernatant.
- Add 0.1–0.5 ml of the culture media to the pellet for swim up.
- Insemination can be done and sperm concentration and motility can be determined using the WHO, 2010 protocol

4.18.4. Swim-Up from Semen

It is recommended for normozoosp semen quality with high sperm count and good motility. Furthermore, it is carried out in a round bottom tube placed at an angle of 45 degrees and this method favors the selection of motile sperm with an intact membrane that results in a higher clinical pregnancy rate in the IVF laboratory.

Method:

- Incubate 4–5 falcon sterile round bottom tube at 37 degrees Celsius for a few hours and add 1–2 ml of sperm washes medium.
- Allow semen to liquefy and carry out semen analysis using the WHO protocol.
- Gently underlay liquefied semen (1–2 ml) in the bottom of the sterile tube.
- Place the round-bottom tube at the 45-degree position in a humidified incubator at 37 degrees or room temperature for 30–60 minutes with tubes tightly capped. The motile sperm cells migrate upward into the sperm wash medium.
- Aspirate the upper and middle sections of the medium in the tubes and combined them in a conical centrifuge tube.
- Centrifuge the pooled swim-up fractions at 300 g for 10 minutes and remove supernatant and repeat centrifugation with an additional 2 ml sperm wash.
- Remove the supernatant.
- Suspend the pellet in an equilibrated bicarbonate-based IVF medium (0.1–0.5 ml) and incubate at 37 degrees in a 5–6% co2 environment for 30 minutes.
- Post-wash analysis is carried out for sperm concentration and motility.
- Use for IVF or ICSI insemination

4.19. MANAGEMENT OF POST-CIRCUMCISION GLANS/PENILE NECROSIS

Male circumcision is among the most commonly performed procedure for traditional, religious, and medical reasons in the world. Circumcision is the process of surgically cutting the foreskin, called the prepuce, covering the glans penis, in a certain shape and length, and exposing the tip of the penis. While it is estimated that 25% of men are circumcised all over the world, this rate is 99% in Turkey and Africa.

Circumcision operations, which have been performed in many societies from the past to the present, are now performed by doctors from various specialties. Although circumcision is not a simple procedure, it is still widely performed by insufficient educated, uneducated, and unlicensed individuals. Recently, with the increasing awareness in society, the number of circumcisions performed by traditional circumcisers tended to decrease and the number of applications made to specialist surgeons increased significantly.

Numerous methods have been portrayed for circumcision activity. The primary motivation behind circumcision is to accomplish a superior corrective outcome. In the writing, the difficulty rate is somewhere in the range of 1 and 15%nafter circumcision.

As per age evaluations, the frequency of confusions after circumcision in babies per age appraisal is accounted for as roughly 0.2-0.6% and this proportion is more than multiple times in the scope of 1 to 9 years. Assuming surgeries for circumcision are performed bit by bit, the difficulties are very interesting. In any case, entanglements, for example, dying, contamination and scar development can, in any case, be seen. Aside from these intricacies, inconveniences, for example, penile corruption, which is a lot more uncommon however hard to treat and make due, can likewise be seen.

The most well-known reasons for penile putrefaction after circumcision incorporate dorsal nerve block with nearby sedative specialists, expanded pressure of the injury dressing, and the utilization of unseemly careful procedures and gadgets.

Even though there are different clinical and careful treatment choices in the writing for patients with penile corruption, there is still no agreement on treatment for the executives. Information in the writing on penile rot couldn't go farther than the case series. The point of this study is to give a medicolegal show of the discoveries and results gained in the assessment of a high volume of cases interestingly.

4.19.1. Circumcision Time

It has been widely reported that as the age of the patient increases, complications of circumcision occur more frequently. Bleeding is more common in the "mini puberty" period of infancy that starts at 4 weeks and extends to 3 months. This is thought to be due to a hormonally-mediated increase in penis and foreskin size and vascularization.

Banieghbal reported only two minor bleeding complications requiring suturing in a recent prospective, observational study of 583 neonatal circumcisions. Both occurred in 3-week-old infants. He reported that, based on the use of the Neonatal Pain Scale, the ideal system for 'painless' circumcision is the first week of life. Horowitz and Gershbein found no complications in 98 babies circumcised with the Gomco clamp in the first month of life and supported neonatal circumcision. However, when they circumcised 3-8 months old children with the Gomco clamp, they found the rate of bleeding requiring suture or fulguration at a rate of 30%

4.19.2. Circumcision Complications

Indications for circumcision generally develop due to medical, religious and cultural reasons. Post-circumcision complications are classified as early and late complications. Early complications are bleeding, pain, inadequate skin excision, and surgical site infection, and these complications are usually treated conservatively or medically. Chordee, iatrogenic hypospadias, penile necrosis and glandular amputation are among the serious early complications.

Late complications include epidermal inclusion cysts, urinary retention, skin bridge formation between the penile shaft and glans, cordial, penile adhesions, phimosis, hidden penis, urethrocutaneous fistula, and meatal strictures. Most of these complications can be avoided by paying attention to surgical principles and specific techniques used.

4.19.3. Glans And Penile Necrosis

Among these complications, glans ischemia or necrosis is rarely seen. Although not all causes of post-circumcision necrosis are presented here, this is multifactorial in the pathogenesis of necrosis, the use of vasoconstrictive agents containing local anesthetic drugs, certain surgical methods, attachment of vascular structures during the procedure, excessive monopolar cautery use, wound bandage compression during nocturnal penile erections and post-circumcision infections are blamed.

Circumcision is a surgical procedure and therefore should only be done by a doctor. Today, it is known that it is done by unlicensed circumcisers in most countries. Complications are more common in circumcisions performed by unlicensed circumcisers. While the complication rate is 1% in circumcisions performed by specialist physicians, it reaches 10% when performed by unlicensed health technicians and reaches 85% when performed in traditional circumcisions.

4.20. EPIGENETICS IN MALE INFERTILITY

Male sterility is an intricate state in which hereditary, epigenetic and environmental ways of life have been distinguished as major contributing factors. In spermatogenesis, which is a complex multistep differentiation process, a large number of mature spermatozoa are every day delivered by the fertile male.

Likewise, this interaction involves an assortment of novel hereditary and epigenetic systems that in the end produce haploid sperm, which gives half of the genetic material and epigenetic data that is expected to make another life upon treatment. The sperm's development is error-prone and could add to male sterility due to chromosomal anamolies at meiosis phase I in the process of spermatogenesis.

It is accepted that epigenetic changes are fundamental to controlling ordinary gonadal turn of events and spermatogenesis. This incorporates the ordinary transmission of variation epi-marks controlling the testis-tissue specific chromatin compaction and the resultant gene expression in a like manner. In this respect, a few lines of proof have featured the presence of unusual epigenetic marks in physical and germ testicular tissues that are related toimpaired fertility or poor semen criterions. .

Hypermethylated genes included in PIWI-related little RNAs (piRNAs) have been seen in testicular tissues of males having different types of issues. Inside this unique situation, a review has announced disturbed epigenetics of cells from various testicular tissues including Leydig and Sertoli cells. Strangely, patients with low testicular volume have been accounted for to have lower chromatin smallness and poor sperm quality.

As of now, endeavors have been focused on understanding the possible key job for epigenetic changes in male reproduction and the inheritance. Epigenetics is characterized as sub-molecular factors or cycles around DNA that control germline movement free of DNA grouping and are mitotically stable.

Epigenetic changes are likewise a group of factors that influence the transmission of genes, however, don't influence the DNA succession. In this part, we analyse the epigenetic control on human sperms quality, that affect the male sterility.

4.20.1. Environment and Epigenetics in Male Infertility

The worldwide climate has changed after some time as an outcome of industrialization and the ever-evolving toxic products. Such toxic products including beauty care products, food , packing materials, toys, and agrochemicals when come in contact with humans affect them badly..

A portion of these toxins appears as endocrine disruptors that could act with others to change the biological equilibriums in normal populations and impact human health and related to an expanded rate of reproductive illnesses.

The development and normal functioning of male reproductive system are thought to be highly sensitive to environmental contaminants exposure/ insults and metabolic status that could adversely affect sperm's number, quality and the reproductive health of the subjected individuals. In line with the thought that the epigenome is more vulnerable than the genome for such environmental insults, a large number of studies have investigated the role of epigenetic modifications in shaping endocrine functions and their potential influence on spermatogenesis.

Due to the protracted period of replication and cell division along the continuous cycles of mitosis and meiosis in adult males spermatogenesis, it is thought that the accumulation of environmentally induced epigenetic are much greater in males than in females.

In addition to studies that highlighted dynamical reaction of sperm epigenome to a wide range of environmental and lifestyle stressors . Sperm epigenome is believed to be affected by a large number of biological factors (including aging, obesity, diet, endocrine disruptors and disease), environmental exposures (such as smoking, alcohol, medications, air pollutions, toxic waste socio-economic stress) and life style (i.e., exercise intervention).

These factors might contribute to primary sources of the increased male factor infertility and decline in seminal parameters. In comparison to the reproductive system's genome, its cellular epigenetic landscape shows a high degree of plasticity, and thus it is more susceptible to be influenced by the environment insults.

Indeed, the different critical timeframes of spermatozoa development represent windows of susceptibility for epigenetic errors to occur and aberrations potentially induced by environmental insults, possibly affecting fertility and embryonic competence. In an attempt to address the question whether genetic predisposition or environment have significant impact on an individual being infertile, earlier monozygotic twins studies have concluded that socioeconomic environment seemed to influence relative magnitude and pattern with certain genetic background.

Given the discoveries that arose out of the factual examination of health review on 1795 Vietnamese male twins , "factors one of a kind" to individual twins could impact their fertility state more conspicuously than external substances, genes or the normal climate impacts.

4.21. CONCLUSION

Thus, it can be said that the entire male reproductive system is dependent on hormones, which are known to be chemicals. These chemicals regulate the activity of many different types of cells or organs. The primary hormones that are involved in the male reproductive system are follicle-stimulating hormone, luteinizing hormone, and testosterone.

Follicle-stimulating hormone is important for sperm production (spermatogenesis), and luteinizing hormone stimulates the production of testosterone, which is also required for the production of sperm. Testosterone is solely responsible for the development of male characteristics, such as muscle mass and strength, fat distribution, bone mass, facial hair growth, voice change, and sex drive.

REFERENCES

1. Bathgate, R., 2022. Male reproductive anatomy and physiology. Manual of animal andrology, [online] pp.10-18. Available at: <https://www.researchgate.net/publication/358494569_Male_reproductive_anatomy_and_physiology> [Accessed 22 June 2022].
2. Koussa, V. and Kuligowska, E., 2007. Anatomy of the Male Reproductive Tract. Male Reproductive Dysfunction, [online] pp.9-20. Available at: <https://www.taylorfrancis.com/books/10.3109/9781420018813-4> [Accessed 22 June 2022].
3. Mega Obukohwo, O., Eze Kingsley, N., Arientare Rume, R. and Victor, E., 2022. The Concept of Male Reproductive Anatomy. Male Reproductive Anatomy, [online] Available at: <https://www.intechopen.com/chapters/79167> [Accessed 22 June 2022].
4. Mjaess, G., Aoun, F., Roumeguère, T. and Albisinni, S., 2022. The Male Reproductive System Anatomy. Men's Health and Wellbeing, [online] pp.1-12. Available at: <https://www.researchgate.net/publication/358913870_The_Male_Reproductive_System_Anatomy?_sg=qZACHG9LfAKxjQII8s_cs37WI6HV-JDmw5rw_oXC36l7-2e-csVlOYUxiFq0eMQyt6bGX_h4rcxjPew> [Accessed 22 June 2022].
5. Toragall, M., C. Ghagane, S., B. Nerli, R. and B. Hiremath, M., 2022. Reproductive Toxicology: An Update. Male Reproductive Anatomy, [online] Available at: <https://www.intechopen.com/chapters/79679> [Accessed 22 June 2022].

CHAPTER 5

PLANT SCIENCE - STRUCTURE, ANATOMY, AND PHYSIOLOGY

CONTENTS

Plant science is the term used for a range of research and scientific studies that actually explore the estimated 410,000 species of land plants found on the earth. The chapter, first talks about the general plant structures, including phloem and its types. Furthermore, anatomy and root development have been explained followed by the roles of phytochromes in the formation of roots. Towards the end, applied plant anatomy is discussed in the context of quality control of medicine.

5.1. INTRODUCTION

The study of plant structure and organs is known as plant science. The word is derived from the Greek words anna, which means up, and tomia, which means cutting. When combined, it means dissection or cutting up. It is regarded by many as one of the oldest medical sciences.

The evolution of anatomy was a result of questions related to religion and philosophy. Early studies were done on animals such as pigs and monkeys before subsequently being done on humans. Alcmaeon, an ancient Greek scientist and philosopher who lived around the 5th century BCE, is believed to be the earliest person known to have pursued the study of anatomy.

Anatomia is the study of the body structures of living things and the identification of their characteristics. Gross anatomy deals primarily with the dissection and observation of major body structures, and it is focused on the human body in its narrowest sense.

"Gross anatomy" customarily refers to the study of those body structures large enough to be examined without the help of magnifying devices, while microscopic anatomy is concerned with the study of structural units small enough to be seen only with a light microscope.

Dissection is basic to all anatomical research. The earliest record of its use was made by the Greeks, and Theophrastus called dissection "anatomy," from ana temnein, meaning "to cut up."

Often we hear the term 'anatomy' while learning about the body and how it works. What is anatomy? What does anatomy mean, how would we define anatomy? The study of anatomies is a specific branch of science that deals with the structure and identification of the human body and its different sections.

Though the phrase "anatomy of the body" is often used about humans and human body parts, it includes all living things. This study of the body structure is divided into two subsections of gross or macroscopic anatomy and microscopic anatomy.

5.2. PHLOEM: CELL TYPES & STRUCTURE

Phloem is the vascular plant tissue responsible for the transport and distribution of sugars produced by photosynthesis. Since the plant is a continuum, phloem will be found in the external part of root cylinders, in the stem vascular bundles, and in the abaxial part of the venations of every single leaf.

Generally, the phloem in roots and stems is external to the xylem and in leaves it is abaxial, but some exceptions are generally taxon-specific. The phloem found on the inside is named internal or intraxylary phloem.

As a constitutive tissue in the plant body, phloem functions extrapolate its main function of sugar transport, including transport of signalizing molecules such as mRNAs, hormones, and defenses from biotic and abiotic agents, sustenance of the organs, gas exchange, and storage of many ergastic materials, such as starch, calcium oxalate crystals, and tannins.

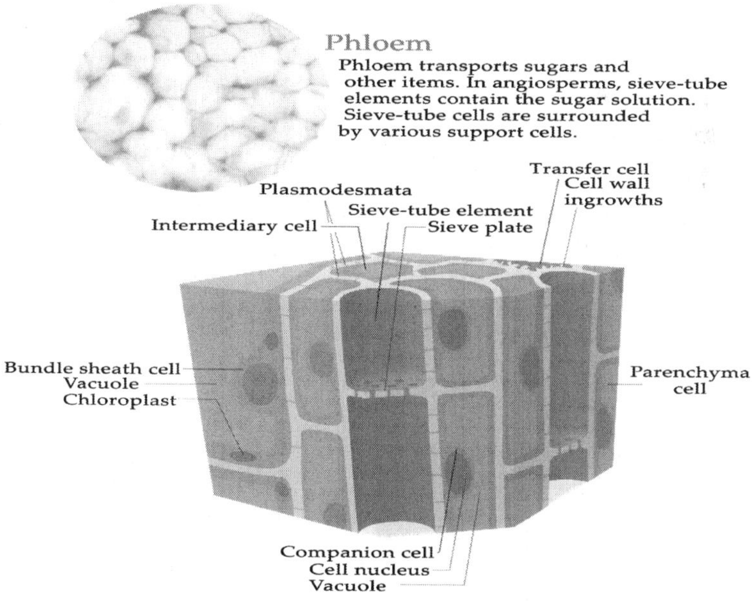

Figure 5.1. Diagram of phloem *tissue in plants.*

Source: Image by Wikimedia Commons

Parenchymatic cells of the phloem can also give rise to new meristems, such as the phellogen or cork cambium. All vascular plants have phloem, which typically includes specialized living conducting cells named sieve elements whose nucleus, ribosomes, and other organelles degenerate during maturation, making sugar transport more efficient.

The life and function of these cells will then rely on closely associated parenchyma cells which support the physiological functions of these sieve elements. Phloem is usually only found in vascular plants, but rudimentary phloem-like conducting cells are found in other lineages as well, including bryophytes, leptoids, and even algae outside the plant kingdom, such as kelps and Phaeophyceae.

In plants with secondary growth, the primary phloem originates from the embryo and the apical meristem procambium throughout life or the cambium.

5.3. PHLOEM CELL TYPES

Phloem is a complex tissue composed of three types of cells, the sieve elements, the parenchyma cells, and the sclerenchyma cells. Sclerenchyma cells might sometimes be absent in primary and/or secondary phloem.

The presence, quantities, and arrangements of these cell types in the tissue commonly vary and may be taxonomic informative. Lists depicting these variations in all phloem cell types are of ultimate importance for complete bark descriptions. What follows is a description of these three major cell types in the phloem.

5.4. CONDUCTING PHLOEM CELLS

The term sieve element refers to all conducting cells in the phloem, including both sieve cells and sieve tubes. The name sieve derives from the strainer appearance given to the cells by the presence of numerous pores crossing their bodies. These pores are specialized plasmodesmata of wider diameter, and the sieve areas are specialized primary pit fields.

Angiosperms have sieve pores that are lined with callose, which was shown to be associated with the sieve pores, but not gymnosperms. Large amounts of callose deposit in the sieve areas also when the sieve element loses conductivity, suffers an injury, or becomes dormant. Callose in gymnosperms is typically wound callose. Callose can be easily detected with aniline blue under fluorescence or resorcin blue.

The walls of sieve elements are only primary walls, but sometimes these walls can be very thick and receive the name of nacreous walls. These walls are present in all major vascular plant lineages. Nacreous walls can be very thick, and some authors have proposed they would be secondary walls.

Nacreous walls can almost occlude the entire lumen of the sieve element; hence, its presence needs to be considered in experiments on sugar translocation. Such thick walls might be related to resistance to high turgor pressures within the sieve elements. Nacreous walls seem to have a strong phylogenetic signal and are much more common in some families, such as Annonaceae, Calycanthaceae, and Magnoliaceae.

There are two types of sieve elements: sieve cells and sieve tube elements. Sieve tube elements feature sieve plates, that is, sieve areas with wider and more abundant sieve pores, usually in both extreme ends of the cells, while sieve cells lack sieve plates. A group of connected sieve tube elements forms a sieve tube. According to this concept, lycophytes and ferns have sieve cells.

Due to the differences in morphology and distribution of protoplasm organelles and chemical substances between gymnosperm sieve elements and vascular cryptogams, Evert suggests that "sieve cell" refers to gymnosperms, while "sieve element" refers to lycophytes and ferns.

The longevity of sieve elements varies. In many species, it is functional for just one growing season, while for other species they can be functional for a couple of years, or in the case of plants that lack secondary growth, they will be living for the entire plant life span. Palm trees would perhaps be the plants with the oldest conducting sieve tube elements, since some reach 200 years.

In other plants, on the other hand, the sieve elements collapse a few cells away from the vascular cambium, corresponding to a fraction of the mm. In a mature tree, most of the secondary phloem will generally be composed of sieve elements no longer conducting.

This region is called non-conducting phloem, in opposition to the area where sieve elements are turgid and conducting, called conducting phloem.

There is no need to use terms such as collapsed and non-collapsed phloem or functional and nonfunctional phloem since there are plants where even non-conducting phloem maintains its sieve elements intact, and although large parts of phloem may not conduct, the tissue as a whole is certainly still functioning in storage, protection, and even dividing or giving rise to new meristems, such as the p.

5.4.1. Sieve Cells and Strasburger Cells

Typically, sieve cells are long and taper at the ends, but they lack sieve plates, meaning that there is no area in the sieve element where the pores are wider. Even though the sieve areas may be more abundant in the terminal parts of the sieve cells, the pores in these terminal areas are of the same diameter as those of the lateral areas of the sieve element.

Sieve cells lack P-protein in all stages of development. The sustenance of the sieve cells is carried by specialized parenchyma cells in close contact with the sieve elements, with numerous plasmodesmata, which maintain the physiological functioning of the sieve cells, including the loading and unloading of photosynthates.

These cells are known either as albuminous cells or Strasburger cells. The name albuminous was initially coined given the proteinaceous appearance of these cells' contents. However, because the high protein content is not always present, the name Strasburger cell, paying tribute to its discoverer Erns Strasburger, is recommended over albuminous cells.

Strasburger cells in the secondary phloem can be either axial parenchyma cells, as is common in Ephedra, or ray parenchyma cells, as is common in the conifers. More commonly, the most conspicuous Strasburger cells in conifers are the marginal ray cells which are elongated and have a larger number of symplastic contacts with the sieve cells.

Sometimes declining axial parenchyma cells also act as Strasburger cells in Pinus. The only reliable character to distinguish a Strasburger cell from an ordinary cell is the presence of conspicuous connections. In the primary phloem, parenchyma cells next to the sieve cells are those which act as Strasburger cells.

5.4.2. Sieve Tube Elements and Companion Cells

In angiosperms, synapomorphy consists of sieve tube elements and companion cells, both of which result from the asymmetrical division of one mother cell. In some instances, these mother cells can divide many times, creating assemblages of sieve tube elements and parenchyma cells to be genetically related.

Sieve tube elements have specialized areas in the terminal parts of the sieve elements in which a sieve plate is present. Within the sieve plate, the pores are much wider than those of the lateral sieve areas, evidencing a specialization of these areas for conduction.

In Cucurbita, the pores in the sieve plate are up to 10 μm in diameter, while the pores in the lateral sieve areas are about 0.1 μm. The protoplast of sieve tube elements contains a specific constitutive protein called P-protein (P from the phloem, also known as slime), which in some taxa (e.g., Leguminosae) is nondispersive and can be seen as coagula inside of the sieve element.

The presence of sieve elements and companion cells is seen in lineages of angiosperms where vessels were lost and tracheids re-evolved, like Winteraceae from Magnoliidae and Trochodendraceae from Eucots, which indicate that these two plant vascular tissues evolved independently from the same meristem initials.

Since the sieve tube element loses its nucleus and ribosomes, the companion cell is the cell responsible for the metabolic life of the sieve elements, including the transport of carbohydrates in and out of the sieve elements.

Companion cells may be arranged in vertical strands, with two to more cells. Other parenchyma cells around the sieve tube integrate with the companion cells and can also act in this matter. Typically, the cells closely related to the sieve tube elements die at the same time as the sieve element loses conductivity.

Sieve tube elements are very morphologically. The sieve plates can be transverse to slightly inclined or very inclined and contain a single sieve area or many. The term simple sieve plate is used when there is only one sieve area, while compound sieve plates are used when there are several sieve areas.

Compound sieve plates typically occur in sieve tube elements with inclined to very inclined sieve plates. In addition, sieve elements with compound sieve plates are typically longer than those with simple sieve plates.

The evolution of sieve elements of both sieve area types has been recorded in certain lineages, such as in Arecaceae, Bignoniaceae, and Leguminosae, and to the present, it is not still clear why the evolution of distinct morphologies would be or not beneficial. The only clear pattern is that compound sieve plates occur in long sieve elements. Phloem that contains many fibers also usually possesses compound sieve plates.

5.5. PARENCHYMA

In the primary phloem, just one type of parenchyma is present and typically intermingles with the sieve elements. Parenchyma is divided into two types in the secondary structure: axial parenchyma and ray parenchyma, derived, respectively, from the fusiform and ray initials of the cambium.

The axial parenchyma in conifers commonly is arranged in concentric, alternating layers. These parenchyma cells contain a lot of phenolic substances, which were viewed as a defense mechanism against bark attackers. In Gnetales, the phloem axial parenchyma appears to be intermingling with the sieve cells. Some of these axial parenchyma cells act as Strasburger cells.

Angiosperms have a more variable distribution of the axial phloem parenchyma, which may appear as background tissue, or it may be arranged in bands and radial rows, or it may be sieve tube-centered.

The distribution of axial phloem parenchyma is commonly related to the abundance of fibers or sclereids. In species with more fibers, it is common to have a more organized arrangement of the parenchyma.

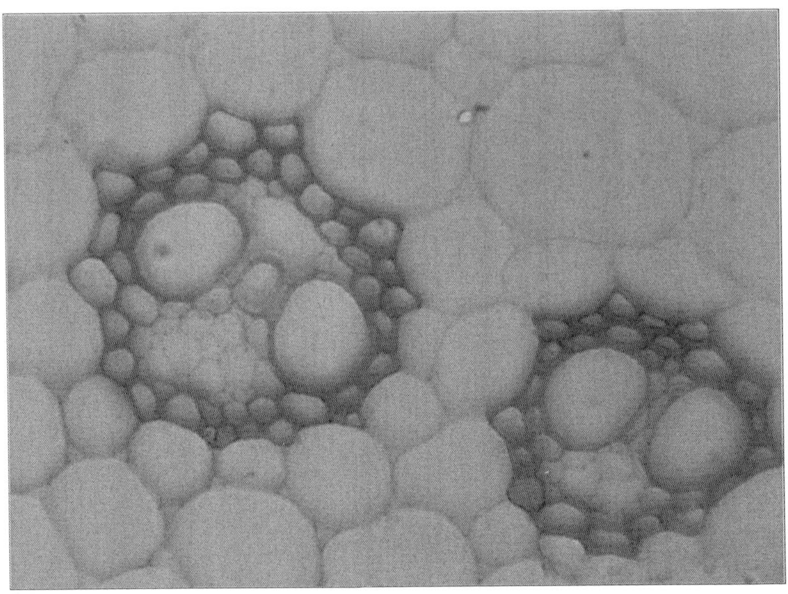

Figure 5.2. Stem Parenchyma.

Source: Image by Wikimedia Commons

For example, in *Robinia pseudoacacia* (Leguminosae) there are parenchyma bands on either side of the concentric fiber bands. When very large quantities of sclerenchyma are present, such as in the secondary phloem of Carya (Juglandaceae) or Fridericia, Tanaecium, Tynanthus, and Xylophragma (Bignoniaceae), the sieve-tube-centric parenchyma appears and, as the name suggests, is surrounding the sieve tubes.

5.6. PHYTOHORMONE-MEDIATED HOMEOSTASIS OF ROOT SYSTEM ARCHITECTURE

Unlike animals, most plants are sessile. It could have been because of this that they developed the ability of organ generation throughout their lives, in contrast to animals, which have a limited ability to generate organs during development.

Half part of the plant body, the root system, is hidden under the ground, where there is a competition of resources, for example, water and nutrients or biotic stresses and abiotic stresses surrounding the root system.

With its strong regeneration ability, the architecture of the root system is shaped by all of these environmental cues together with the internal developmental signals. In this process, phytohormones work as the regulatory molecules mediating the internal and external developmental signals, thus controlling the morphology and function of the root system architecture.

5.7. ANATOMY AND DEVELOPMENT OF ROOT

5.7.1. Root System Architecture

In different plant species, root system architecture (RSA) has diverse morphologies. Two types of RSA exist, the taproot system (or allorhizic system) in gymnosperms and dicotyledons, like *Arabidopsis thaliana* (Arabidopsis), tomato (*Solanum lycopersicum* L.), carrot (*Daucus carota*), and poplar (*Populus* spp.) and the fibrous root system (or homorhizic system) in monocotyledons likes maize.

The taproot system consists of a single thick central primary root (PR) with thin or no lateral roots (LRs); the fibrous root system has small and short-lived primary and adventitious roots (ARs) derived from shoots, stems, or leaves

5.7.2. Intrinsic Developmental Signals and Environmental Conditions Modify Root System Architecture

Arabidopsis as a eudicot has a taproot system, which consists of an embryonic radicle-derived PR and postembryonic-developed LRs and ARs. Root regeneration exists throughout the plants' lifetime; it is a distinctive feature of plants and contributes to their robustness in adverse conditions.

In Arabidopsis, LRs initiate from pairs of pericycle cells that possess developmental potential as plant stem cells. These pericycle cells are selected and directed to become LR founder cells and form LRs by both intrinsic and environmental signals.

The root meristem initiates the initial LR, where root cap-derived auxin influences oscillatory gene expression in the basal meristem and the elongation zone of the root, leading to the pre-patterning of LR initiation sites. The pre-patterning process is marked by the expression of a series of genes, like GATA23, MEMBRANE-ASSOCIATED KINASE REGULATOR 4 (MAKR4), and IAA19.

In the basal meristem and elongation zone, DR5: Luciferase expression was observed to rhythmically pulse with a period of ~6 h, which matched with the period of LR pre-branch site production. It is recently reported that the source of auxin is provided by the cyclic programmed cell death of root cap cells.

It is noteworthy that not all of the pre-branch sites emerge to be LRs. When water availability, nutrients, physical obstacles, or damage are present at the dormant pre-branch sites, LR formation may occur as a selective mechanism. Interestingly, many of the external signals converge on phytohormones to regulate root development. Among these phytohormones, auxin functions as a central mediator.

Mechanical forces are important regulators of plant morphogenesis. LRs always emerge from the convex side of PR bending, resulting in a left-right alternation of LRs. Bending caused by gravitropic curvature led to the initiation of LRs, where subcellular relocalization of PIN1 was observed.

Release of the pericycle cells from the restraints of adjacent endodermis by targeted single-cell ablation of endodermal cells triggered the pericycle to reenter the cell cycle and induced auxin-dependent LR initiation. Excision of the Arabidopsis PR leads to the promotion of LR formation, which is mediated by activated auxin biosynthesis and auxin transport

5.8. ROLES OF PHYTOHORMONES ON ROOT FORMATION

5.8.1. Auxin

At the same time, auxin, a phytohormone that is essential to many aspects of plant growth and development, is also known to play a role in LR development. The natural auxin, indole-3-acetic acid (IAA), is mainly synthesized in a two-step pathway from tryptophan.

First, tryptophan is converted to indole-3-pyruvate (IPA) by the TAA1/SAV3 family of aminotransferases; IPA is then converted to IAA by the YUCCA (YUC) family of flavin monooxygenases. Auxin biosynthesis has been shown to play an essential role in both programmed and wound-induced LR and AR developments.

The polar auxin transport (PAT), which is mediated by auxin influx (AUX1 and LAXs), generates auxin gradients and maintains a maximum auxin concentration to control LR formation and positioning.

Auxin signaling is known to be an integrator of endogenous and exogenous signals for root branching. It begins with the degradation of a class of AUXIN/INDOLE-3-ACETIC ACID (Aux/IAA) through Transport Inhibitor Response 1 (TIR1) auxin receptor, resulting in the activation of the Auxin Response Factor (ARF).

ARF7 and ARF19 transcription factors further induce the expression of downstream target genes like Lateral Organ Boundaries Domain/Asymmetric Leaves 2-LIKE (LBD/ASL) family genes LBD16/ASL18 and LBD29/ASL16, promoting LR initiation at the protoxylem-pole pericycle cells

5.8.2. Cytokinin

Cytokinin is also the main player in root development. In higher plants, isopentenyl adenine (iP), trans-zeatin (tZ), and dihydrozeatin (dZ) are the predominant cytokinins. Cytokinin levels and patterning in the plant are controlled by a fine equilibrium between cytokinin synthesis and catabolism.

Cytokinin biosynthesis is dependent on the activity of ATP/ADP-isopentenyl transferase (IPT) and LONELY GUY (LOG) gene family, and the metabolism is mainly through the Cytokinin Oxidase/Dehydrogenase (CKX) genes.

Cytokinin can also be inactivated through conjugation with glucose. The specific expression of genes responsible for cytokinin synthesis and catabolism is responsible for the spatial distribution of cytokinin.

In Arabidopsis, Histidine Kinase (AHK), AHK2, AHK3, and AHK4/WOL1/CRE1 are transmembrane receptors that detect cytokinin and activate the type-A and type-B Arabidopsis Response Regulators (ARRs) that negatively and positively regulate cytokinin signaling, respectively.

Two type-A ARRs, ARR7, and ARR15 were induced by both cytokine and auxin and are essential for embryonic root development.

Although some evidence showed that cytokinins act as both local and long-distance signals, and some transporter proteins are involved in cytokine transport, the molecular mechanisms of cytokinin transport are still not well characterized.

This postembryonic root development is regulated by the root apical meristem (RAM), where cytokinin acts antagonistically with auxin to regulate cell division in the division zone and cell differentiation in the transition zone, which underpins the maintenance of the RAM and affects the growth and patterning of roots.

Application of cytokinin reduces the number of meristem cells and the size of RAM and promotes cell differentiation in the transition zone; cytokinin biosynthesis and signaling mutants as well as CKX overexpression mutants have a larger RAM with more meristem cells.

Conversely, auxin treatment increases meristem size and promotes cell division in the proximal meristem, and auxin transporter PIN mutants display a smaller RAM.

The cross-talk of cytokine and auxin in regulating RAM activity was shown to converge on the auxin-inducible AUX/IAA family gene SHORT HYPOCOTYL 2/INDOLE-3-ACETIC ACID 3 (SHY2/IAA3) in the transition zone, where cytokinin activates SHY2 via the AHK3/ARR1 two-component signaling pathway to suppress PIN3 and PIN7 expression and promote cell differentiation, while auxin suppressed SHY2 protein, leading to the activation of PINs and promotion of cell division.

Furthermore, Moubayidin et al. revealed that in the transition zone, SCR, a member of the GRAS family of transcription factors, directly represses the expression of ARR1, which controls auxin production via the ASB1 gene and sustains stem cell activity, to simultaneously control stem cell division and differentiation and ensure coherent root growth.

Cytokinin affects the expression of multiple PINs differentially in a tissue-specific manner to regulate auxin distribution. Auxin-cytokinin interactions lead to the generation of distinct hormonal response zones, thus controlling the development of root vascular tissue.

Contrary to auxin, which is a positive regulator of LR development, cytokinin acts as a negative regulator of LR formation. Cytokinin suppresses LR initiation through downregulating PIN expression and preventing the establishment of auxin gradient in LR founder cells.

Mutants with reduced cytokinin levels or deficient cytokinin signaling increased the number of LRs, while cytokinin treatment suppresses LR initiation and development. Li et al. reported that cytokinins inhibit LR initiation by blocking the cell cycle of the pericycle founder cell at G2 to M transition phase while promoting LR elongation by stimulation of the G1 to S transition.

Chang et al. showed, via mutant analysis, that cytokinin biosynthesis genes IPT3 and IPT5 and each of the cytokinin receptor genes AHK2, AHK3, and CRE1/AHK4 act redundantly at the early stages of LR initiation, and lateral root primordia (LRP) formation is particularly cytokinin-sensitive. They suggest that cytokinin may serve as a positional signal for new LRP formation.

In rice, ERF3 interacts with WOX11 to promote crown root initiation and elongation by regulating the cytokinin-responsive gene RR2. Cytokinin has also been shown to modulate LR formation by mediating environmental cues. Jeon et al. showed that Cytokinin Response Factor 2 (CRF2) and CRF3 encoding APETALA2 transcription factors regulate Arabidopsis LR initiation under cold stress.

5.8.3. Other Phytohormones

Other phytohormones, like abscisic acid (ABA), gibberellic acid (GA), brassinosteroid (BR), jasmonic acid (JA), ethylene, and strigolactone (SL), also participate in root growth and development.

5.9. EFFECT OF PHYTOHORMONES ON SHOOT REGENERATION IN RICE CALLUS CULTURE

Phytohormones are considered one of the major important factors that control cell fate in callus cultures. Different types and concentrations of auxin treatment may induce different tissue differentiation. Besides, ABA is also considered to play roles in somatic embryogenesis and shoot organogenesis. The functions of auxin and ABA in regenerable callus induction is known.

5.9.1. Roles of Auxin in Regenerable Callus Induction

Auxin is known to play a major role in cell elongation, growth tropisms, and cell fate determination. Among the endogenous auxin compounds, indole-3-acetic acid (IAA) is the main one, which can be synthesized either through tryptophan-dependent or -independent pathways.

After being synthesized in apical meristems, auxin will be transported to the other tissues through its transporter, PIN-formed proteins (PIN), and AUXIN1/LIKE-AUX1 (AUX/LAX) proteins to deliver the hormonal signal for downstream auxin responses.

The distribution of auxin in plants is polarity which means that it is not equally expressed in the whole plant but specifically concentrated in some tissues or limited to cells. Due to this property, scientists can monitor the patterns of auxin gradients during plant development.

From the literature, the positions of maximum auxin accumulation were the place for processing organ initiation. Therefore, auxin is usually found to have maximum levels in apical meristems or the developing tissues.

During embryogenesis, the expression pattern of PINs dynamically changes within the developmental stages. For example, PIN1 is expressed without polarity until the 16-cell stage, but later in the 32-cell stage, it will express specifically in the basal part of provascular cells to direct auxin flow to hypophysis.

After dividing into transition and early heart embryo stages, the expression of PIN1 will shift to the flank of the apical part, thus accumulating auxin in the edge of the apical domain to shape the embryo. Similar developmental patterns can be found during shoot apical meristem (SAM) establishment, where auxin is highly concentrated in leaf primordia and the central region of SAM.

In callus, the signal of auxin is mainly located in the central region near the callus induction medium. Later, the signal will shift to the surface layer and start SAM establishment for shoot differentiation after transfer into the regeneration medium.

However, this organogenesis process is blocked when the activity of auxin transporters is inhibited, and so does in auxin sensor shoot meristems (STM) mutants. In previous studies, we compared endogenous IAA levels between NRC and HRC in rice callus and found that HRC has higher IAA content than NRC, and so do the mRNA expression levels of PIN1, suggesting that HRC has higher auxin sensitivity, which results in higher regeneration ability.

Also, there are research articles mentioning that overexpression of STM can maintain the somatic embryogenesis frequency even under low concentrations of 2, 4-D. Besides, increasing the expression levels of auxin biosynthesis regulator YUCCA (YUC) genes can also promote shoot organogenesis ability in the callus.

Similar results can be found in the supplement of IAA precursor anthranilic acid (An) in rice callus culture; the shoot regeneration frequency increased by 35% under A treatment, suggesting that a high endogenous auxin level is required for HRC formation.

5.9.2. Roles of ABA on Regenerable Callus Induction

ABA is usually considered to play a negative role in plant growth. In addition to regulating seed dormancy and stress responses, ABA is also implicated in developing roots and shoots.

Although there were some genes participating in ABA signaling, which were also reported to be involved in the shoot regeneration process, there is still no clarity about the molecular function of endogenous ABA on the shoot organogenesis process.

However, some studies found that ABA was highly accumulated in HRC, but not in NRC. Furthermore, the expression profiles of ABA biosynthesis genes were also upregulated, which matches our results.

On the other hand, ABA is reported to stimulate dehydration responses during embryo maturation stages. From the publications, we observed that HRC has less water content and smaller callus size, which is similar to the dehydration phenotype.

Thus, it is possible that ABA shared similar regulation mechanisms with embryo maturation during shoot regeneration.

Despite the loss of water in HRC, it also showed higher content of soluble sugars. It has already been demonstrated that high sugar content enhances osmotic stress and then stimulates endogenous ABA synthesis to cause ABA responses, but the underlying mechanisms of shoot differentiation have thus far remained unclear.

5.10. CROSS TALK AMONG OSMOTIC STRESS AND PHYTOHORMONES IN CALLUS CULTURE

The cause of osmotic stress in cells could come from a lack of water or a large amount of salt, which creates an imbalance between plastids and apoplasts. To achieve tolerance to osmotic stress conditions, cells will modulate the content's osmotic adjustments like sugars, potassium ions, or proline to balance their osmotic pressure to environments to avoid collapse.

It is known that phytohormone ABA accumulates under osmotic stress and modulates anion channel SLAC1 for stomatal closure and potassium transporter KUP for potassium homeostasis, and that osmotic stress can help embryonic callus formation in plant tissue culture.

It was found that HRC showed dosage responses to osmotic treatments and has the highest induction frequency under 0.6 M sorbitol treatment. The osmotic requirements for rice species vary as well. Some cultivars require higher osmotic stresses for HRC formation and others require lower levels, and even one cultivar forms HRC without osmotic treatments.

The ABA response in those calli showed that HRC does have higher Late Embryogenesis Abundance 1 (LEA1) gene expressions, which is commonly used as an ABA signaling marker.

Interestingly, LEA1 is also upregulated in the rice cultivar without osmotic stress treatment, indicating the effect of osmotic stress could be on stimulating ABA biosynthesis and its downstream responses during embryonic callus induction. Not only ABA but auxin is also regulated by osmotic stress. In rice, the endogenous levels of auxin are reported to be suppressed under osmotic stress.

However, a closer look at the expression levels of different auxin biosynthesis-related genes and the distributions of auxin showed various patterns in the whole plants in different stages, some of them even inconsistent with the overall patterns, suggesting that auxin may function differently among tissues under osmotic stress.

As for the patterns of PINs under osmotic stress, one of the researchers reported that osmotic stress may inhibit the expression of PIN1 in leaf primordia, thus suppressing shoot development.

However, previous works indicated that OsPIN1 could be upregulated by 0.6 M sorbitol treatment in HRC, and also in the nonosmotic requirement cultivars. Although there have been many studies performing transcriptomic or proteomic studies of auxin responses under osmotic stress, how osmotic stress directs auxin responses to HRC formation is still less known.

5.11. ROLES OF CARBOHYDRATE METABOLISMS DURING HRC INDUCTION UNDER OSMOTIC STRESS TREATMENT

Exogenous carbohydrates are used either as energy sources or osmotic agents in tissue culture. There have been many studies demonstrating that different carbon sources may result in a different ability to induce callus, but very few have discussed the carbohydrate metabolism and signaling pathways in callus cultures.

Sucrose is widely used as the main carbon source as well as an osmotic agent in rice tissue culture. During the tissue culture, sucrose will be taken up and hydrolyzed into glucose and fructose by CINs, or transported into cells by SUTs for further application. CINs were already reported to be involved in early seeding establishments and grain fillings, and so do the SUTs.

According to some studies, HRCs exposed to osmotic stress were found to contain greater amounts of glucose, sucrose, and starch than NRCs. This was also reflected in higher dry weights for HRCs induced by osmotic stress. The expression patterns of CIN1 and SUTs were analyzed during rice callus culture.

The expression of CIN1, SUT1, and SUT2 in HRC was upregulated by osmotic stress, but not in NRC, while in the nonosmotic requirement cultivars, there these sucrose-uptake genes were expressed earlier than in the cultivar of low-regenerable ability. It is suggested that higher soluble sugars in HRC may be caused by higher sucrose uptake and translocation under stress treatment.

Besides, it was also observed that osmotic stress-induced HRC has lower α-amylase activities and thus increases the content of starches, while in the nonosmotic requirement cultivar; the callus rather induced ADPG pyrophosphorylase (AGPase) activity to accumulate starches. The results suggested two different regulations on carbohydrate metabolisms to HRC induction.

Although there are different ways of starch accumulation in HRC, the accumulated carbohydrates will soon degrade after transplanting the callus onto the regeneration medium in 3 days, suggesting that these carbohydrates could be stored as a carbon source in HRC and then used for the developmental process during the regeneration stages.

Plants are known to accumulate starch granules specifically in the columella cells. Although the function of these starch granules is mostly reported in root gravity, they could also be markers to point out the stem cell niche, since these starch granules may disappear in the plants with stem cell defects. Our studies also found the accumulation of starch granules in peripheral regions in HRC.

A high concentration of sorbitol or mannitol will enlarge the distribution of starch granules. It may link to the increase of the shoot organogenesis area. However, the physiological functions and mechanisms of starch accumulation remained, requiring further studies.

Further, AnA treatment (anthranilic acid and ABA supplements into a medium together) can be used to stimulate the accumulation and metabolism of soluble sugars and starch instead of osmotic stress. High levels of endogenous IAA and ABA at the same time are necessary during HRC induction.

Both of them need to decrease suddenly in a few days and are also important criteria for further shoot regeneration. To link these metabolic changes with phytohormone regulations, we also introduced auxin transport inhibitor TIBA during callus induction and found that it will inhibit carbohydrate accumulation and result in low shoot regeneration frequency.

However, ABA signals seemed to be promoted under TIBA treatment. It is still not clear whether exceeding ABA signals will turn into a negative regulator on HRC induction, but these results still indicate that there must be interactions among auxin, ABA, and carbohydrate metabolisms on HRC formation.

5.12. PHYTOHORMONES

In nature, phytohormones are substances that are naturally produced in low concentrations and play an important role in the regulation and expression of genes. They are a diverse group of signaling molecules that result in a variety of cellular and developmental processes, and signaling networks in plants under biotic and abiotic stress.

They work as chemical messengers to communicate cellular activities and act either at their site of synthesis or elsewhere in plants following their transport to higher plants.

Different phytohormones interact with each other and show synergistic or antagonist interactions that might be helpful in tolerance mechanisms. A large number of phytohormones are studied to date; among them, jasmonates are the emerging players in environmental stress tolerance.

5.13. JASMONATE: A POTENT PHYTOHORMONE

"Jasmonic acid and its methyl ester methyl jasmonic acid are cyclopentanone phytohormones prepared from oxidized lipids (oxylipins), which originate from α -linolenic acids (α -LAs). "

First isolated JA from the culture filtrate of the fungus *Lasiodiplodia (Botryodiplodia) theobromae*, a plant pathogen, was identified as a plant growth inhibitor, whereas its derivative methyl jasmonate was first isolated from *Jasminum grandiflorum* (jasmine) petal extract. (+)-7-iso-Jasmonoyl-L-isoleucine (JA-Ile) is the best-described bioactive JA to date, but other JAs like cis-jasmone, jasmonoyl ACC (JA-ACC), and jasmonoyl isoleucine (JA-Ile) are also studied by scientists with multiple biological functions.

Various developmental and environmental factors are responsible for the production of JA in membranes and resulted in the expression of stress-tolerant genes. JA is ubiquitously found in the plant kingdom and results in the expression of genes at the transcriptional and post-transcriptional levels.

An imperative role of jasmonate (when applied exogenously in low concentrations) is reported in enhanced pathogen resistance.

The interaction of Jasmonates with various phytohormones, such as ABA, salicylic acid, ethylene, etc., results in diverse developmental processes such as seed germination, seedling growth, pollen fertility, fruit ripening, senescence, and tolerance. However, its extent of effectiveness entirely depends on the type of plant species tested or its concentration.

MeJA is more volatile than JA, so exposure to it either in solution or in the gaseous phase can elicit plant responses. Apart from its significant role in plants, derivatives of jasmonates, e.g., methyl jasmonate is used as a fragrant constituent in many aromatic mixtures.

5.13.1 Biosynthesis of Jasmonates: The Octadecanoid Pathway

In 1984, the biosynthesis pathway of jasmonates was identified. The scientists Vick and Zimmerman were the first to illustrate the biosynthetic pathway in a simplified manner which indicated that linolenic acid could be converted into cyclopentanone 12-oxo-phytodienoic acid (12-oxo-PDA) through lipoxygenase enzyme.

Methodically the pathway was studied in model plants like Arabidopsis and tomato. The octadecanoid pathway of Jasmonates completes in two cellular organelles such as chloroplasts and peroxisomes and are considered to be the primary sites. α-Linolenic acid (α-LeA) released from galactolipids (due to wounding or pathogens attack) of chloroplast membranes is found to be the main player in MeJA and JA production.

Phospholipases1 (PL1), lipoxygenase, allene oxide synthase, and allene oxide cyclase (AOC) are significant enzymes involved in the biosynthesis of jasmonates.

5.13.2. Role of Enzymes Involved in Jasmonates Biosynthetic Pathway

- **Phospholipases**: formation of α-LeA from chloroplastic membrane lipids
- **13-Lipoxygenase (LOX):** addition of oxygen molecules to α-LeA and results in the formation of an intermediate compound 13-hydroperoxy-9, 11, 15-octadecatrienoic acid (13-HPOT).
- **Allene oxide synthase (AOS**): oxidation of 13-HPOT to allene oxide
- **Allene oxide cyclase (AOC):** formation of 12-oxo-phytodienoic acid (12-OPDA) an unstable compound from allene oxide.

The AOS and AOC are present in plastids and they act in concert. As a biosynthetic pathway product, 12-OPDA is formed in chloroplasts and undergoes three cycles of 3-oxidation in peroxisomes.

A methylation reaction by JA methyl transferase produces jasmonate methyl ester, which is methyl jasmonate. Among the six 13-LOXs of Arabidopsis, four of them are (LOX2, LOX3, LOX4, and LOX6) but LOX2 is a vital lipoxygenase in JA biosynthesis.

5.13.3 Emerging Roles of Jasmonates

5.13.3.1. Inhibitory Action on Seedling Growth

Exogenous application of JA possibly shows inhibition of primary root growth, leaf expansion, and hypocotyl elongation which ultimately leads to inhibitory action in seedling growth. InsP5 enhances the interaction of COI1 with JAZ9 and the inhibitory effect of JAs on root growth.

JA represses leaf expansion by inhibiting the activity of the mitotic cyclin CycB1; 2 and cell division, rather than by affecting cell size.

Transcription factor-like MYC2 and its close homologs show both positive and negative effects on hypocotyls in red/far-red light and blue light conditions; it works positively in inhibition of hypocotyl elongation in red/far-red light and negatively regulates the inhibition of hypocotyl elongation by blue light. ERF109 binds to and activates anthranilate synthase A1 (ASA1) and YUCCA2 (promoters of the auxin biosynthetic genes) and results in the promotion of lateral root formation in Arabidopsis.

5.13.3.2. Role in Plant Reproductive Development

Various transcription factors of the R2R3-MYB family-like MYB21, MYB24, and MYB57 are direct targets of JAZ proteins. These TFs have a significant role in mediating JA-regulated stamen development. Formation of MYB-MYC complexes due to the association of MYB21 and MYB24 with the IIIe bHLH factors MYC2, MYC3, MYC4, and MYC5 controls stamen development in Arabidopsis.

Besides the role of JA on stamen development, JA plays a major role in seed and embryo development in tomatoes the jasmonic acid-insensitive1 (jai1) mutant, which exhibits a loss of function of the tomato homolog of COI1, cannot set viable seeds.

Moreover, the production of OPDA and a residual amount of JA, in tomato mutant acx1a, set viable seeds. Gene silencing (OPR3 silenced gene) in SiOPR3 a transgenic line of tomato produces a comparable amount of OPDA to wild type and sets only a few viable seeds; methyl-JA treatment can restore the seed setting of SiOPR3.This further suggests the role of methyl jasmonate in maternal control of seed development.

5.13.3.3 Role in Abiotic Stress Tolerance

As a result of abiotic stresses like drought, salinity, high and low temperatures, heavy metal toxicity, and others, plant morphological, physiological, biochemical, and molecular changes occur; such stresses impair plant growth and productivity. JA is believed to play a role in plant responses to abiotic stresses including drought, salt, and heat stress.

Salt is one of the hazards that causes physiological drought by delaying seed germination, and seedling establishment, and reducing the growth and yield of any crop. Under salt stress, jasmonates proved to be an imperative phytohormone in mitigation.

Jasmonates recovered salt inhibition on dry mass production in rice and diminished the inhibitory effect of NaCl on the rate of 14CO2 fixation, and protein content in *Pisum sativum*.

The pleiotropic effects of MeJA in protecting plants have been reported for several plants, reported in his studies JA is responsible for the amelioration of chilling injury, water stress, and salinity stress in *Oryza sativa* L., *Lycopersicon esculentum* L., *Fragaria vesca*, and *Hordeum vulgare*.

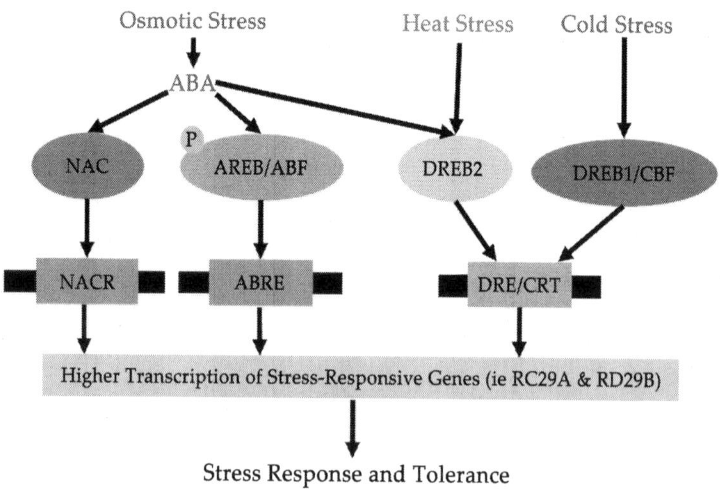

Figure 5.3. A general pathway of TF regulation in relation to abiotic stress.

Source: Image by Wikimedia Commons

High-temperature stress destructively influences plant processes and disturbs cell homeostasis. Heat shock proteins (HSPs) are synthesized in plants in response to high temperatures that prevent denaturation and assist refolding of damaged proteins. Electrolyte leakage assays in heat-stressed plants after application of low-concentration MeJA demonstrate the cell viability responses.

Heating WT Arabidopsis led to the accumulation of several jasmonates including OPDA, MeJA, JA, and JA-Ile, and the expression of jasmonate inducible gene PDF1.2 was found to be high upon heat stress exposure.

The suppressor of the G2 allele of SKP1 (SGT1) protein operates as a cofactor of heat shock protein 90 (HSP90) in both plants and mammals forming functional complexes and providing thermotolerance. JA also showed an essential role in heavy metal and nutrient toxicity.

In a study, they reported that the excess amounts of boron present in soil decrease the net photosynthetic rate, the closing of stomata, internal CO2 concentration, and total chlorophyll content in leaves. Foliar application of boron-stressed plants of Artemisia with MeJA started to stimulate the synthesis of antioxidant enzymes, reduce the amount of lipid peroxidation, and enhance artemisinin content.

Researchers have discovered that jasmonate can both kill cells and suppress cell proliferation, as noted in the first report on anticancer activities caused by jasmonate. MJ was studied in topical application for precancerous and cancerous skin lesions

5.13.3.4. Role in Biotic Stress Tolerance

Two types of responses are shown by plants due to tissue injury, and jasmonates play a significant role in these responses by signaling in plants. Two types of responses are local response and systemic response.

In local response during tissue damage, various attacker-derived signals or damaged-associated plant-derived signals are produced; these signals are either chemical or physical and are recognized by PRRs pattern recognition receptors present on the cell surface.

This recognition event activates de novo synthesis of JA and JA-Ile by an unknown pathway. SCF COI1/26 proteasome activation by JA-Ile results in the degradation of JAZ proteins. These proteins are responsible for the repression of transcription factors (TFs) involved in the expression of defense-related traits.

5.14. A REGULATORY CIRCUIT INTEGRATING STRESS-INDUCED WITH NATURAL LEAF SENESCENCE

The onset of leaf senescence is a highly regulated developmental program that is controlled by both genetics and the environment. Multiple stresses in plants induce programmed cell death, and the underlying regulatory mechanisms are often associated with molecular links of developmentally programmed senescence.

The transcriptome changes induced by different environmental stressors are not entirely overlapping, but functional analysis of genes commonly induced as shared responses can give clues on signaling integration.

This approach has been used to select overlapping genes as candidate regulatory components that integrate the ER stress and osmotic stress responses, which were shown later to participate also in natural leaf senescence.

Among genes identified as components of the ER and osmotic stress shared response, the developmental and cell death (DCD) domain-containing asparagine-rich proteins (NRP-A and NRP-B) were the first ones to be characterized as cell death-promoting proteins, and hence this multiple stress-integrating signaling was designated as stress-induced DCD/NRP-mediated cell death response.

Further characterization of the cell death pathway implicated in the discovery of the signaling module ERD15/NRPs/GmNAC81:GmNAC30/VPE also has been shown to operate in developmentally programmed leaf senescence.

5.15. MODEST OVERLAPPING OF ER STRESS AND OSMOTIC STRESS RESPONSE IDENTIFIES NRPS AND NACS AS CELL DEATH-PROMOTING GENES

5.15.1. Osmotic Stress Responses

As a result, organisms are constantly adapting to their environment, which activates sensor proteins for forwarding transmission to downstream effectors responsible for assembling adaptive cellular responses. Abiotic stresses consist of a set of adverse environmental conditions that limits plant development.

Cold, high temperature, salinity, water availability (drought or floods), radiation, pollution, and chemical exposure are the most common examples of types of abiotic stresses

Generally, a signaling sensor network connects internal and external stimuli to adaptive responses leading to molecular modifications that allow physiological adjustments, which ultimately cause susceptibility or tolerance to the exposed conditions.

Molecular responses to abiotic stress conditions in plants are crucial for survival and productivity as these stresses often limit yield. Among abiotic stresses, drought and excess salinity conditions induce sophisticated adaptive responses in plants to cope with or acclimate to these adverse environmental conditions.

Some types of abiotic stress responses are better understood than others. In plants, for example, the molecular mechanisms of perception and responses to drought, high salinity, and endoplasmic reticulum stress are well characterized, and many stress-related cell signaling pathways are completely elucidated, revealing some convergence points between them.

The osmotic stress in plants, caused by water deprivation or high salinity, for example, undergoes a set of characteristic morphological, molecular, and physiological changes. One of the most notorious symptoms in plants under low water availability is the ABA-mediated stomatal closure.

This hormone-mediated morphological change affects plant physiology. Closing the stomata reduces evapotranspiration, maximizing the usage of water by the cell, but it also causes a reduction in carbon dioxide uptake. This imbalance in photosynthetic activity results in ROS (reactive oxygen species).

5.15.2. ER Stress Responses

The endoplasmic reticulum (ER) is one of the most dynamic organelles in cell machinery. It is the gateway for the synthesis of secretory proteins and contains the necessary apparatus to ensure quality protein synthesis, protein maturation, and secretion in eukaryotic cells.

Furthermore, the ER can modulate some chronic stress-related pathways, promoting oxidative stress, autophagy, and apoptotic cell death in mammals and plant cells.

Several adverse environmental conditions can affect the ER quality control machinery, causing unfolded/misfolded protein accumulation in the ER lumen. The secretory proteins are synthesized in ER membrane-bound polysomes, and, as soon as they enter the organelle, they are processed by the ER processing machinery.

Under normal conditions, there is a perfect balance between the rate of protein synthesis and ER processing capacity. Any conditions that disrupt this balance promote unfolded/misfolded protein accumulation in the ER lumen. Consequently, the disturbance of ER function triggers a complex and coordinated signal cascade, which is detected by ER membrane-associated sensors.

These sensors activate the expression of ER-resident chaperones, foldases, and ER quality control components. Collectively, these cytoprotective mechanisms are known as the unfolded protein response pathway.

The detection of ER stress is mediated by membrane-associated sensors, identified both in mammals and plants. In mammals, there are three of these sensors: kinase/endoribonuclease inositol-requiring enzyme 1 (IRE1), activating transcription factor 6 (ATF6), and protein kinase RNA-like ER kinase (PERK), which are regulated by the ER-resident molecular chaperone BiP (binding protein).

The ER sensors initiate the UPR to restore ER homeostasis under stress conditions. If the adverse physiological status is prolonged, they can initiate some alternative routes leading to cell death.

5.15.3. Convergence of ER Stress and Osmotic Stress Responses into a Cell Death Signaling Pathway

Under physiological conditions, UPR involves three protective mechanisms: translation suppression by PERK-mediated IF2α phosphorylation, the upregulation of molecular chaperones in the ER, and proteasomal-mediated protein degradation by the ERAD pathway.

However, if the stress conditions are sustained and the UPR pathway fails to restore ER homeostasis, apoptotic pathways are triggered as an ultimate attempt to survive. In plants, there is a specific branch of ER stress that integrates the osmotic stress and leads to programmed cell death (PCD), the development and cell death domain-containing N-rich protein (DCD/NRP)-mediated cell death signaling.

This cell death pathway was first identified via genome-wide and expression profiling approaches, which revealed a modest overlapping between ER and osmotic stress-induced transcriptomes of soybean seedlings treated with PEG (an osmotic stress inducer) and tunicamycin and AZC (ER stress inducers).

Several genes displayed similar kinetics and a synergistic induction under combined ER and osmotic stresses, indicating that the ER stress response integrates the osmotic signal to potentiate transcription of shared target genes.

Among them, are two plant-specific DCD/N-rich proteins, NRP-A and NRP-B, a ubiquitin-associated protein homolog (UBA), and a NAC domain-containing protein, GmNAC81, which displayed the most robust synergistic upregulation by the combination of both stresses.

Transient expression of NRPs or GmNAC81 in soybean protoplasts and *Nicotiana benthamiana* leaves demonstrated that they are critical mediators of ER stress- and osmotic stress-induced cell death in plants.

5.16. EARLY DEHYDRATION RESPONSIVE GENE 15, ERD15-LIKE, CONTROLS NRP EXPRESSION

The early dehydration responsive (ERD) genes were first identified due to their rapid induction in response to drought stress. The ERD genes (ERD1 to ERD16) encode a set of proteins that differ in biological functions and cellular localization.

Among them, ERD15 is a small acidic and hydrophilic protein that belongs to the PAM2 domain-containing protein family. The PAM2 domain is a well-characterized protein-protein interaction domain, which allows ERD15 to interact with polyA-binding proteins (PABP) regulating mRNA stability and protein translation.

In addition to PAM2, ERD15 contains two other domains with unknown functions, designated as PAM2-associated element 1 (PAE1) and QPR.

ERD15 is a multiple stress-responsive gene that contributes to the adaptation to abiotic and biotic stress. Light treatment, cold stress, and high salinity trigger ERD15 expression. ERD15 functions as a negative regulator of the abscisic acid (ABA)-mediated response and a positive regulator of the salicylic acid (SA)-dependent defense pathway.

ERD15-overexpressing transgenic lines are less sensitive to ABA and display an enhanced salicylic acid-dependent defense pathway, which was associated with increased resistance to the bacterial *Erwinia carotovora* of the transgenic lines.

5.17. THE STRESS-INDUCED NRP/NAC081/VPE MODULE TRANSDUCES A CELL DEATH SIGNAL

Cell death induced by ER stress and osmotic stress signals is mediated by the DCD/NPR signaling pathway, in which NRPs and GmNAC81 play a critical role. More recent progress toward deciphering this branch of stress-induced cell death signaling includes the identification of two additional downstream components, the NAC transcriptional factor (GmNAC30) and the vacuolar processing enzyme (VPE).

GmNAC30 was identified as a nuclear partner of GmNAC81 via two-hybrid screening using GmNAC81 as bait. GmNAC30 and GmNAC81 exhibit similar expression profiles and cell death activity. They are upregulated by ER stress, osmotic stress, and the cell death-inducer cycloheximide.

Consistently, GmNAC30 promotes cell death when transiently expressed in soybean protoplasts and, as a downstream component of the cell death signaling, is induced by the expression of NRP-A and NRP-B.

Both GmNAC30 and GmNAC81 interact with each other *in vitro* and *in vivo*. The complex formed binds to common cis-regulatory sequences within target promoters and stimulates hydrolytic enzyme promoters in plants, including the caspase-1-like vacuolar processing enzyme (VPE) gene that is involved in PCD.

Consistent with their transcriptional function as a heterodimer, GmNAC81, and GmNAC30 display overlapping and coordinated expression profiles in response to multiple environmental and developmental stimuli. Therefore, the stress-induced GmNAC30 cooperates with GmNAC81 to activate PCD through the upregulation of the cell death executioner VPE.

5.18. A NEGATIVE REGULATOR OFTHENRP/ NAC081/VPE SIGNALING MODULE CONFERS TOLERANCE TO DROUGHT

Stress conditions can negatively modulate the NRP/DCD-mediated cell death response of plants. Moreover, this modulation improves cellular stability and consequently increases the plant tolerance to stress conditions in an essential process that is required for plant acclimatization and development.

The molecular chaperone BiP plays a crucial role as a negative regulator of NRP/DCD-mediated cell death response. BiP belongs to the HSP70 family, which is essential to protect the cells against environmental stresses and restore cell homeostasis.

The molecular chaperone BiP has a catalytic site at the amino-terminal region and a substrate-binding site at the carboxy-terminal region. BiP is involved in the regulation of several processes in the endoplasmic reticulum, a critical organelle that is related to responses to abiotic and biotic stress in plants.

In the ER, BiP acts as a sensor that responds to quantitative and qualitative changes in the ER by regulating the activity of ER stress transducers. Furthermore, BiP coordinates the cell death signaling, which connects the signals from osmotic and ER stress in a DCD/NRP-dependent manner.

BiP attenuates the NRP/DCD-mediated cell death signal propagation by the modulation of expression and activity of the pathway signaling components. BiP overexpression in soybean attenuates ER stress- and osmotic stress-mediated cell death, a phenotype that is linked to a delay in the induction of GmNRP-A, GmNRP-B, and GmNAC81 under ER stress and osmotic stress.

Furthermore, enhanced accumulation of BiP in tobacco (*Nicotiana tabacum*) prevents the GmNRP- and GmNAC81-mediated induction of cell death-associated physiological and molecular markers, whereas silencing of endogenous BiP enhances the cell death response.

5.19. THE STRESS-INDUCED DCD/NRP-MEDIATED CELL DEATH SIGNALING POSITIVELY REGULATES LEAF SENESCENCE

The senescence of leaves begins with a physiological transition between active photosynthetic leaves and degenerative and nutrient-recycling leaves. The classical age senescence-related symptom is the leaf yellowing caused by generalized chlorophyll loss.

Figure 5.4. *Heracleum sphondylium* leaves showing chlorophyll loss.

Source: Image by Wikimedia Commons

The age-induced senescence or naturally programmed leaf senescence, hereafter referred to as leaf senescence, occurs by plant aging and is precisely regulated by senescence-associated genes (SAGs).

Many SAGs are environmental- and stress-responsive genes, integrating a convergent regulatory cascade between natural plant development and stress-induced PCD.

At the molecular level, the onset of senescence is accompanied by a massive reprogramming of gene expression, probably controlled by senescence-associated transcription factors. Among these, several NAC transcription factors have been associated with senescence regulation based on high-resolution temporal expression profiles.

5.20. MEDICINAL PLANTS: THEIR PARTS, USES, AND ECOLOGY REVIEWED

Herbal remedies available in nature are used by traditional Ethiopian healers to treat a variety of diseases. As reported before, approximately 800 species of medicinal plants grown in Ethiopia are used for treating about 300 medical conditions. However, based on the present review, the number of medicinal plants and the treatments/medications identified and listed are limited as presented here section by section.

5.21. MEDICINAL PLANTS AND THEIR GROWTH FORMS AND PARTS USED

5.21.1. Composition and Growth Forms of Medicinal Plants

As reported by many authors, there are different types of medicinal plant species with their parts, habitats, and disease types being treated. There are 80 medicinal plant species with 63 genera, used by the local communities for various human treatments.

There are a number of common medicinal plants that are commonly used to treat and cure various diseases, including *Aloe* species, *Eucalyptus globulus*, *Hagenia abyssinica*, *Cupressus macrocarpa*, *Buddleja polystachya*, *Acmella caulirhiza*, *Citrus* species, *Clematis* species, *Coffee arabica*, *Croton macrostachyus*, *Euphorbia* species, *Ficus sycomorus*, and *Moring*.

Figure 5.5. *Ficus sycomorus* fruits.

Source: Image by Wikimedia Commons

Based on the review, all plant growth forms were not equally used as remedies, because of the difference in distribution among the growth forms. Accordingly, the life forms of medicinal plants reviewed were 18 trees (22.78%), 23 shrubs (29.11%), 29 herbs (36.71%), 3 climbers (3.81%), 4 trees/shrubs (5.06%), and 2 herbs/shrubs (2.53%). Of all life forms, herbs were, thus, the major medicinal plants used by the community for human treatment followed by shrubs and trees.

5.21.2. Medicinal Plant Parts Used for the Preparation of Traditional Remedies

According to the review, traditional healers use plant parts for preparing medications that are variable in nature. Healers mostly used fresh specimens from commonly available plants to prepare remedies for their patients; this might be mostly due to the effectiveness of fresh medicinal plant parts in treatment since the contents are not lost before use compared to the dried ones.

As also referred by many authors, the traditional healers have harvested leaves, roots, barks, seeds, fruits, stems, flowers, barks, seeds, or latex of medicinal plants to prepare their traditional medicines for their patient treatments. Most remedies were prepared from the leaf (32.98%) and root (29.79%) parts of the medicinal plants to treat the diseases compared to the other parts of them.

Figure 5.6. Traditional Medicine.

Source: Image by Wikimedia Commons

This finding of the review is in line with the findings of the majority of authors' papers. The main reason that many traditional medicine practitioners used the leaf parts compared to others for remedial preparation is due to their accessibility and to prevent them from extinction.

Harvesting the root parts of the medicinal plant for the preparation of traditional medicines has negative consequences on the existence of the plants themselves in the future. That is why most medicinal plants are currently at risk, declining highly due to them using their root parts besides other human pressures.

5.21.3. Uses of Medicinal Plants in Treating Different Disease Types

Using these medicinal plants revised, the local communities could be able to treat about 69 disease types. These medicinal plants were used to treat skin diseases, gonorrhea, diarrhea, wounds, tapeworms, snake bites, stomachaches, headaches, evil eye, heartburn, cancer/tumor, and malaria.

Particularly, most of the patients (who come from rural areas) with their prospective disease types have been treated by traditional healers, before coming to clinics and/or hospitals located far away many kilometers from their residential areas.

The disease types most frequently treated by traditional medications (traditional healers) provided by those medicinal plants were stomachaches, wounds, cancers/tumors, skin diseases, headaches, toothaches, coughs, and diarrhea, which took the first, second, third, fourth, fifth, sixth, and seventh ranks, respectively, although the majority of disease types were frequently treated less than four times, ranging from one to three times.

Additionally, it shows that a single medicinal plant species can treat more than one disease type.

It is for this reason that medicinal plants play a significant role in providing traditional medicine, prepared by local healers, and thus being used to treat and cure diseases that affect the local communities where they occur.

Even, following the traditional uses and effectiveness of the medicinal plants, the traditional healers are also popular in the local societies, providing cultural values. The study also confirmed that the traditional health practitioners are with a good knowledge of medicinal plants used to treat different diseases of their locals.

5.22. ECOLOGY AND/OR HABITATS OF MEDICINAL PLANTS

As referred from the revised documents for this review, the habitat preference of medicinal plants varied from place to place. Many medicinal plants can be found near rivers, streams, and wetlands, disturbed sites, grasslands, cultivated lands, woodlands, bushland, and home gardens.

Generally, the majority of medicinal plants were found in the wild compared to those plants found in cultivated and home gardens together. Many of the authors of the reviewed articles confirmed that the majority of medicinal plants were collected from natural habitats or wild by traditional practitioners compared to home gardens.

Among medicinal plants found along stream/riverbanks, the majority of them are supposed to be medicinal plants having herbal life forms/habits. This could be due to their shallow roots, which cannot bring water from the deep parts of their habitats.

Most medicinal plants have also been lost due to human activities such as over-harvesting, fires/deforestation, agricultural expansion, overgrazing, and urbanization. This implies that the availability and accessibility of most medicinal plants in Ethiopia are also very difficult.

Hence, most of the medicinal plants were restricted to areas (such as cliffs, hills/mountains, gorges, disturbed areas, riverbanks, and valleys of the wild) that are not easily accessible to use/harvest.

Not only is this, but also the knowledge of traditional practitioners pertinent to the identification of medicinal plants with their parts and ecology and the process of preparation of herbal medicines and medication with their quality/effectiveness is declined/lost since the knowledge is mostly transferred orally from generation to generation, not documented.

In this way, human effects on the natural habitat of medicinal plants pose a threat to the conservation of medicinal plants and the knowledge associated with them. In Ethiopia, medicinal plants and their ethnobotanical knowledge are under serious threat as a result of ecological and socioeconomic changes.

5.23. APPLIED PLANT ANATOMY: QUALITY CONTROL OF HERBAL MEDICINE

Plant materials are used throughout developed and developing countries as home remedies, over-the-counter drug products, and raw materials for the pharmaceutical industry, which represents a substantial proportion of the global drug market. Thus, traditional herbal medicines and their preparations have been widely used for thousands of years in many countries.

Thus, it is important to describe some modern control histological techniques or tests, appropriate standards, and examples of how to assess the quality of medicinal materials and their products.

Quality control of herbal medicine using histological techniques and pharmaceutical practices is also very vital for avoiding the risks happening to patients and the beliefs in services provided by traditional healers.

According to, quality control is a phrase that refers to processes involved in maintaining the quality or validity of the manufactured products.

However, the quality control of herbal medicine is beyond this, meaning it is the management of medicinal plants and their products during cultivation, identification process of the plant species with their parts and localities (there being free from the polluted environment causing diseases), and medicine preparation including its components, medication processes, storage standards, and dosage; all should be taken into account.

This means that without proper all-around quality control, there is no assurance that the contents of the herbs contained in the package are the same as what is stated outside the package.

Climatic factors (prevailing temperature, rainfall, humidity, altitude of the growing region, light), nutritional factors (nutrients, pH, cation exchange capacity), harvesting factors (age, season, collection time, plant organ), and post-harvesting factors (storage hygiene, drying process) are the major factors affecting the contents and composition of medicinal plant raw materials and their products.

Several important laboratory test methods (histological techniques), as well as good pharmaceutical practices, are used. Several techniques for evaluating the quality of herbal drugs and meat are widely used, including thin-layer chromatography, microscopic analysis, and electrophoresis.

These techniques and good pharmaceutical practices are also used to support the development of national standards based on local market conditions, with due regard to existing national legislation and national and regional norms.

Consequently, improved and currently available pharmaceutical analytical methods have improved harvesting schedules, cultivation techniques, storage, product purity, and the activity and stability of active compounds.

REFERENCES

1. 2020. Plant Science - Structure, Anatomy and Physiology in Plants Cultured in Vivo and in Vitro.
2. Bazira, P., 2022. Anatomy of the kidney and ureter. Surgery (Oxford), [online] Available at: <https://www.surgeryjournal.co.uk/article/S0263-9319(22)00103-X/fulltext> [Accessed 13 June 2022].
3. Cunningham, A., 2022. 'The anatomy of the blood'. 'I Follow Aristotle': How William Harvey Discovered the Circulation of the Blood, [online] pp.127-139. Available at: <https://www.researchgate.net/publication/360501676_%27The_anatomy_of_the_blood%27> [Accessed 13 June 2022].
4. Laitman, J., 2022. The anatomy of antisemitism and antisemitism in anatomy: A personal reflection. The Anatomical Record, [online] 305(4), pp.849-859. Available at: <https://www.researchgate.net/journal/The-Anatomical-Record-Advances-in-Integrative-Anatomy-and-Evolutionary-Biology-1932-8494> [Accessed 13 June 2022].
5. Thesen, T., 2022. Anatomy. Composition for the 21st ½ century, [online] pp.141-186. Available at: <https://www.researchgate.net/publication/359807407_Anatomy> [Accessed 13 June 2022].

CHAPTER 6

VETERINARY ANATOMY AND PHYSIOLOGY

CONTENTS

It is important to have a sound knowledge of anatomy and physiology for the effective clinical treatment of companion animals. This field has diverse areas to be explored. The chapter talks about the history and some imaging technology being used, initially. Later, the anatomy, histology and physiology of a hoof have been discussed. Later, myocardial metabolism has also been detailed. Towards the end of the chapter, the role of veterinarians in medicine conservation and animal welfare has been discussed for a better understanding of the reader.

6.1. HISTORY OF VETERINARY ANATOMY AND PHYSIOLOGY

People have long been captivated by animal anatomy, with mural paintings illustrating the superficial anatomy of animals reaching back to the Paleolithic age. Nevertheless, evidence shows that the first appearance of scientific anatomical research may have occurred in ancient Babylonia, albeit the tablets on which this was written have perished, and the remnants show that Babylonian knowledge was rather restricted.

As a result, ancient Egyptian civilization is thought to be the birth of the anatomist, with early research of anatomy chronicled in the writing of numerous papyri. The Edwin Smith papyrus, which dates to 3000 BCE, shows recognition of cerebrospinal fluid, meninges, and surface anatomy of the brain, but the Ebers papyrus depicts the systemic function of the body, including the heart and vascular, gynecology, and tumors.

The Ebers papyrus is estimated to have been created about 1500 BCE, although it is also likely to be based on previous writings. A school of anatomy was formed in Alexandria in the early third century, and it was the first of its type to perform human dissection. It was established in the Greek system and adhered to Hippocratic beliefs.

Hippocrates (c. 460–370 BCE) characterized the human mind as being separated into two halves by a thin vertical membrane, as it did in other animals. Herophilus and Erasistratus were the most famous Alexandrian doctors. Many of their students traveled and practiced all throughout the Mediterranean region.

Many notable Greek anatomists were educated at Alexandria, but then as Egypt began to collapse, advancements in Greek philosophical and scientific culture started to outstrip all contemporary knowledge of human anatomy and physiology. In roughly 500 BCE, the early scientist Alcmaeon of Croton, a Greek doctor, is commonly credited for attributing the mind to the brain.

He also noticed that the arteries and veins in his animal dissections looked different. Herophilus advanced these notions around the brain, identifying it as the center of the neurological system. Herophilus was born around 335 BCE in Alexandria, was educated there, and stayed there during the reigns of the first two Ptolemaios Pharaohs. He is credited as being the first anatomist to execute a methodical dissection of the human body.

According to Roman Marcus Aurelius (161–180 CE), physicians were more likely to maim or kill their patients since they were unaware of the anatomy necessary. Marcus was well-known for his research efforts to better grasp the anatomy and physiology of animals ranging from monkeys and snakes to cattle and cats, both adults and children.

Marcus even saw parallels between macaques and humans. Through the dissections of many organisms, especially mammals, reptiles, and insects, the Greek Aristotle eventually became renowned as the founder of comparative anatomy.

As the second century began, the Greek anatomist Galen established a variety of anatomical and physiological ideas, primarily through animal dissection. Galen was the discoverer of the recurrent laryngeal nerve and had numerous sophisticated wound healing ideas. The work influenced anatomical and physiological research for the next fifteen centuries, allowing substantial advances in medicine.

Most of the subsequent anatomical research, up until about the eleventh century, was therapeutically focused and accomplished in response to the documented works of early authors such as Galen. Depending on the work with human corpse dissection, anatomist Mondino dei Luzzi's emergence in the thirteenth century prompted development in anatomical and physiological study.

While Andreas Vesalius, a Renaissance figure, advanced physiological studies with realistic explanations of the mechanics of pulmonary breathing, among many other revelations in his 1543 treatise entitled 'De Humani Corporis Fabrica,'.

William Harvey, born in England in 1578, was the first one to accurately describe the circulatory system, distinguishing blood flow returning to the heart and the heart operating as a muscle pump. Harvey also hypothesized the concept of capillary anastomoses connecting the artery and venous systems, which were not confirmed until after his death in 1661 by Marcello Malpighi.

When William Hunter encouraged pupils to do their own dissections on cadavers in the eighteenth and nineteenth centuries, dissection became a more important part of anatomical research. However, this resulted in 'body-snatching,' and as a result, The Anatomy Act (1832) was enacted to restrict the purchase of human cadaveric material. Such a rule favored hospital-based anatomy schools; it was noted that science was becoming more professional.

Anatomy and physiology are unquestionably key components of medical education today, and may be taught in a variety of ways, including dissection,' self-directed learning,' and 'problem-based learning.' Recent technological advancements have enabled the incorporation of digital anatomical models into university courses, offering the modern student greater access to the study of anatomy.

Figure 6.1. Museum of Veterinary Anatomy.

Source: Image by Wikimedia Commons

6.2. IMAGING TECHNOLOGY WITHIN ANATOMY AND PHYSIOLOGY

Fewer instruments and procedures for visualizing the body were accessible in the days of the early anatomists and physiologists, as stated above. Many approaches have been created by scientists throughout the years that are vital for veterinary and human anatomists and physiologists alike. Dissection and sketching have long been important skills that are being employed today.

As the first microscopes were constructed, the capacity to view through tissue and cells enhanced both anatomy and physiology, as did medical practice. The first modern microscope, invented in 1590 by Hans and Zacharias Janssen, has undoubtedly evolved over time. There is further evidence to imply that early microscopes and lens magnification were used and ideas developed in early China 4000 years ago, as well as by the ancient Greeks and Romans.

Hooke also created a microscope and penned the now-famous 'Micrographia,' and Antonie van Leeuwenhoek, the Father of the Microscope, developed this work and had his Royal Society papers verified by Hooke. The contemporary microscope has improved significantly, with optical microscopy, which uses a nonlinear optical phenomenon, electron microscopy, confocal microscopy, and even hand-held microscopes being accessible for anatomical, physiological, and medical study and diagnosis.

Since the advent of X-rays, advances in medical imaging have given a vital tool for study, offering unprecedented viewing of the body. The combination of postmortem cadaveric dissection with live patient imaging methods has shown to be a powerful tool in studying the anatomy and physiology of the living animal.

Wilhelm Roentgen, a German scientist, saw X-rays for the first time in 1895. Roentgen deduced that bone and metal were opaque on radiographs by imaging his wife's hand, and medical applications of the discovery swiftly followed. Marie Curie saw bone injuries in troops on French battlefields using portable X-ray equipment. While the use of X-ray crystallography was critical in deciphering the genetic code, which in turn had a huge influence on physiology.

Godfrey Hounsfield later broadened the use of X-rays by creating computer software that could combine several radiological pictures to provide a three-dimensional vision inside the body. This was the invention of computed tomography (CT). CT had a substantial advantage over radiography alone in that it allowed for the visualization of diverse soft tissue types. The first CT scan of a patient was performed in 1971, successfully scanning the brain for a tumor in the frontal lobe. Because CT has important practical applications, Hounsfield was given the Nobel Prize in Physiology or Medicine in 1979.

The hazards of utilizing ionizing radiation were recognized, particularly when imaging the fetus, and as a result, X-rays were reduced and replaced by ultrasonography and magnetic resonance imaging. In a 1958 publication, Ian Donald pioneered the use of ultrasonography in obstetrics and gynecology. Since then, two-dimensional ultrasound technology has advanced tremendously, and three-dimensional ultrasound can now map and measure blood flow. Ultrasound has been a watershed moment in medical imaging and a foundational approach to non-invasive research.

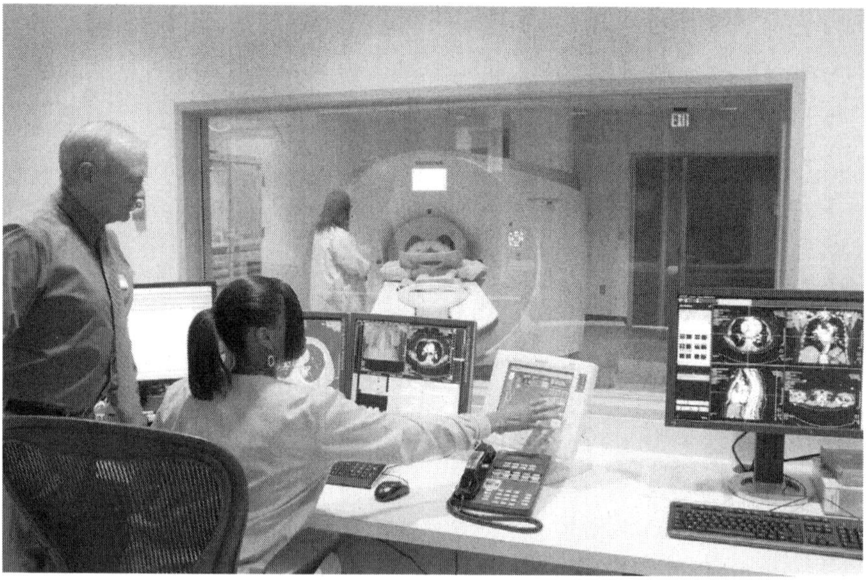

Figure 6.2. Radiology Imaging.

Source: Image by Flickr

Yet, in recent years, magnetic resonance imaging (MRI) has become a critical diagnostic and research tool. Felix Bloch and Edward Purcell discovered nuclear magnetic resonance in 1946, laying the groundwork for the creation of contemporary MRI. Paul Lauterbur invented NMR by using gradients to magnetic fields to form a two-dimensional picture, while Sir Peter Mansfield developed procedures for/slice selection, image creation, and interpretation.

These breakthroughs led to the creation of MRI as we know it today, and the men were awarded the Nobel Prize in Medicine or Physiology in 2003. MRI is a vital medical imaging method that uses no ionizing radiation and provides a viable alternative to invasive treatments.

Imaging mass spectrometry is a new advancement in imaging that allows tissue samples to be seen on a molecular level without the need for chemicals or antibodies. A group of scientists, including Caprioli, invented the mass spectrometry imaging method, which is particularly sensitive for usage on proteins and peptides. This method, however, cannot map the transcriptome, and as a result, a new approach, mass spectrometric imaging, has been created.

6.3. WOMEN IN VETERINARY MEDICINE, ANATOMY AND PHYSIOLOGY

In 1762, Claude Bourgelat established the first veterinary school in Lyon. The Veterinary College in London, the United Kingdom's first veterinary school, did not open until 1791. There were no standards governing the study of veterinary medicine at the time, and practitioners sometimes lacked official training. Fifty years later, the Royal Charter of 1844 authorized the establishment of the Royal College of Veterinary Surgeons, granting the profession official legitimacy.

The Veterinary Surgeons Act of 1881 differentiated qualified practitioners from untrained practitioners. The veterinary practice was originally focused on horses, with many veterinarians working in the army and public sector. When engines became less popular and horses became less in demand, the number of equine veterinarians decreased, and the profession began to focus on farm animals and livestock. Interest in companion animal care surged in the mid-twentieth century, as did the small animal industry.

Women's participation in anatomy and physiology research has increased throughout time. These women have backgrounds in veterinary medicine, and the fundamental sciences (biology, anatomy, and animal science), or have transferred from human medicine and anatomy to veterinary research. Aleen Cust, the first female veterinary surgeon in the United Kingdom, graduated from Edinburgh Veterinary College in 1900.

The Royal College of Veterinary Surgeons declined her membership at the time, therefore Cust was unable to receive a certificate from the RCVS. Cust did not receive full RCVS membership until 1922, thanks to The Sex Disqualification (Removal) Act of 1919, which prohibited gender discrimination.

Mary Brancker was the first female president of the British Veterinary Association and a founding member of the Society of Women Veterinary Surgeons. Brancker had enormous power and was awarded the OBE for her efforts during the 1967–1968 foot and mouth disease pandemic. In addition, she was granted the CBE in 2000 for her commitment to animal health and welfare. Women's role in veterinary medicine grew further when Dame Olga Uvarov became the first woman president of the RCVS in 1976.

Although being a formerly male-dominated vocation, the veterinary profession has experienced significant growth in the number of female veterinary surgeons in recent years. Gender equality, as embodied in the Women's Education Act of 1974 in the United Kingdom and analogous initiatives and policies across the world, has been critical in permitting females to obtain admission to veterinary institutions.

In 2006, 51 percent of working veterinary surgeons in the UK were female; by 2014, this figure had risen to 57.6 percent. Women currently account for over 80% of veterinary medicine and science graduates in numerous countries, including the United Kingdom, Slovenia, and the United States of America, but there is still considered to be a disparity in the sorts of fields in which women later choose to work.

Males are deterred from entering the field owing to the increasing number of women, according to the decline in the number of men applying to veterinary institutions. Although there are several historical reasons why fewer women entered the industry, it is obvious that the number of women joining the field is growing globally.

6.4. THE ANATOMY, HISTOLOGY AND PHYSIOLOGY OF THE HEALTHY AND LAME EQUINE HOOF

The histo-morphological intricacy of internal hoof structures appears to restrict effective research of the horse foot. One of the most devastating pathological conditions of the horse foot is foot lameness. Foot lameness has historically been associated with hoof deformity in most species, and a collection of molecular processes has been established in relation to the condition.

So far, there has been debate on the prevalence of foot lameness in horses, as it is unclear if foot lameness causes hoof deformities or vice versa. Knowing the fundamental anatomy of the horse foot is critical for furthering research into the structures' participation in the development of lameness and for a better understanding of illnesses such as lameness and laminitis. This chapter seeks to demonstrate the anatomy and physiology of the hoof and bones of the equine foot, as well as how they relate to lameness and laminitis in horses.

6.4.1. Gross Anatomy of the Equine Hoof

The distal extremities of domestic mammals are wrapped by a keratinized capsule, which takes the shape of a hoof capsule in ungulates and a claw in carnivores. This insensitive horny structure encloses the distal part of the second phalanx (also known as the middle phalanx or short pastern bone), the distal phalanx (also known as the coffin bone or the pedal bone), and the navicular bone, as well as connective tissues such as the distal interphalangeal joint, medial and lateral hoof cartilage, as well as the terminal end of the deep digital flexor tendon and the navicular bursa.. These components are linked together to form a cohesive and durable structure inside the foot.

The hoof is made of horn, which is formed from epidermal tissue that has been keratinized to varying degrees. Horn is organized primarily as a collection of parallel tiny tubules joined by the intertubular horn. This structure is important for weight carrying since it surrounds practically the whole circle of the foot, folding inwards towards the back to produce the bars that support the heels.

The sole encases the palmar/plantar surface of the foot and is concave, with a comparable but softer and more flexible structure than the hoof wall. The hoof connects to the skin at the coronet, where it is covered by the periople, a waterproof ring of the soft tubular horn. The white line, which is very elastic and generated from the epidermal lamellae, connects the sole's perimeter to the hoof wall.

The frog is an elastic structure composed of a pliable, incompletely keratinized horn that is crucial for shock absorption, blood circulation, and slip prevention. The frog continues inwards to the digital cushion, which is made up of poorly vascularised adipose tissue embedded in a fibroblastic mesh and is involved in shock absorption as well as blood pumping. The existence of the distal digital annular ligament separates the digital cushion from the deep digital flexor tendon (DDFT).

The equine foot bones include the third phalanx (P3; sometimes known as the distal phalanx), the second phalanx (P2), and the navicular bone. P3, commonly known as the pedal or coffin bone, is the main bone of the foot, occupying the most distal location and connecting to the hoof capsule via the lamellar and solar coria.

P3 helps to strengthen and stabilize the hoof capsule and is very porous due to the abundance of nutrient foramina. P2, or the short pastern, forms the proximal interphalangeal, or pastern, joint with P1 and the distal interphalangeal, or coffin, joint with P3. P2's short, virtually cuboidal composition makes it resistant to a wide variety of pressures.

The navicular, also known as the distal sesamoid bone, is a tiny, smooth bone found at the distal interphalangeal joint caudal of the spine. It has a pulley-like function, coated ventrally in smooth fibrocartilage, that allows the DDFT to glide easily under the distal interphalangeal joint without interference from adjacent bones. The navicular synovial bursa and distal synovial sheaths help the DDFT move smoothly by secreting synovial fluid that lubricates the region.

The superficial digital flexor tendon (SDFT) is part of the back-tendon pair, along with the DDFT, which descends from the deep digital flexor muscle in the forearm to the flexor surface of P3. The SDFT joins to the proximal surfaces of P1 and P2 after descending from the superficial digital flexor muscle in the forearm.

The common digital extensor tendon oversees interphalangeal joint extension (CDET also known as m. extensor digitorum communis in the fore limb and m. extensor digitorum longus on the rear limb). The CDET descends the leg dorsally from the long digital extensor muscle anterior to the knee, finishing at the extensor process of P3 with projections into P1 and P2.

While the DDFT and SDFT allow the interphalangeal joints of the foot to flex and the CDET allows them to extend, the existence of lateral and medial collateral ligaments restricts the joints' adduction and abduction, respectively. The collateral ligaments attach to notches on the distal and proximal edges of P1 and P2 for the proximal interphalangeal joint, and to notches on the distal and proximal edges of P2 and P3 for the distal interphalangeal joint.

Three pairs of chondral ligaments link to the medial and lateral cartilages of P3 to keep it in place. The navicular suspensory ligaments anchor to the distal border of P1, immediately dorsal to the collateral ligament attachments, then converge at the navicular bone, producing the distal navicular ligament, which ends at P3.

The coria are the densely vascularized and innervated dermal areas that lie between and support the skeletal components as well as the epidermal hoof capsule. The coronary corium runs along the proximal border of the hoof wall, with each tubule forming around short, finger-like papillae extending from the coronary corium, which supply the proliferative epidermal cells and keep the hoof expanding.

The solar corium is structurally and functionally identical to the coronary corium, with papillae allowing the sole to develop. sensitive or dermal lamellae of the lamellar corium constitute, together with the epidermal/insensitive lamellae of the inner hoof wall with which they interlock, the suspensory apparatus of the third phalanx, suspending P3 within the hoof capsule. Each dermal lamella has a pair of papillae, the terminal papillae, which produce the soft, elastic white line that connects the wall to the sole.

6.4.2. Vasculature of the Foot

The vasculature, in addition to the bones, ligaments, tendons, and other soft tissues of the foot, is crucial in the horse foot. The blood vessel system transports dissolved gases, nutrition, waste, signaling molecules like hormones, and immune cells to and from other organs. The extensive circulatory network runs beneath the hoof capsule and through the bones, nourished by branches from the medial and lateral digital arteries and returned to general circulation via the medial and lateral digital veins.

The common palmar digital artery and the dorsometatarsal artery furnish the vascular blood supply of the hoof, with these primary branches giving birth to medial and lateral palmar/plantar digital arteries. The tiny plantar common digital arteries contribute to the formation of the digital arteries in the hind limb. There are branches at the level of the second phalanx that supply the heel bulbs and coronary area.

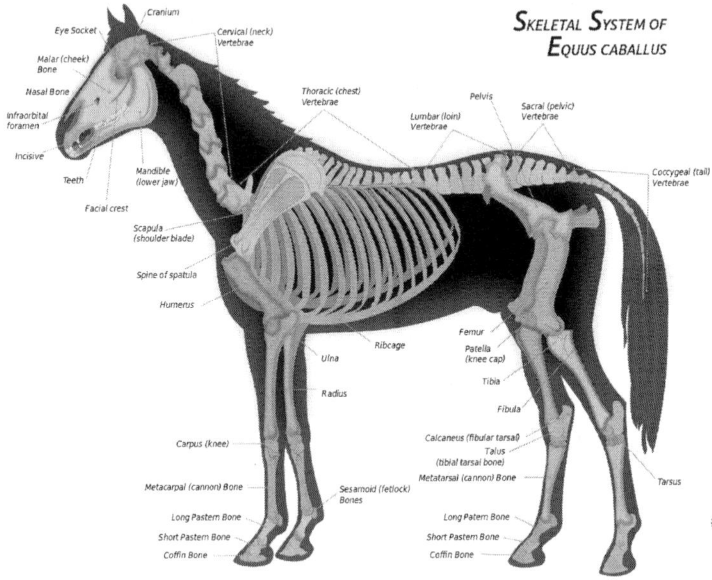

Figure 6.3. Horse Limb Anatomy.

Source: Image by Wikimedia Commons

The dermal vascular arteries are divided into three independent arterial blood supplies: the dorsal coronary corium, the palmar/plantar portion of the coronary corium and laminar corium, and the dorsal laminar corium and solar corium, as blood flow is reversely directed from the distal to the proximal part of the dermal lamina (also referred to as lamella/lamellae and lamellar in the literature).

The blood supply's terminal branches enter the distal phalanx from the medial and lateral sides, forming multiple anastomoses within the bone to create the terminal arch. There are 8–10 blood vessels arising distally from this arch to supply the sole edge. This is a critical blood vessel organization in horse feet because the terminal arch and its branches are covered by the bony canal, which can be changed in chronic laminitis, resulting in ischemia and a reduction in the corium's development rate.

Because the dorsal and palmar sections of the foot have separate blood supply and drainage paths, the capillary network of the horse digit is complicated. The blood veins of the dorsal lamella, for example, run via the distal phalanx and the blood supply of these regions is directed from distal to proximal, whereas the palmar lamella is from proximal circumflex to dorsal lamella.

Thus, as in the case of the founder, bleeding from the sublamellar circulation might result in the rotation of P3. As a result, the arteries from the plexus have thicker walls and smaller lumens, and are unable to self-regulate the volume fluctuations involved in smooth muscle contraction as well as encompassing arteriovenous shunts.

The horse hoof veins are classified into three types based on their location: wall dermis veins (proximal and distal areas), coronary dermis veins, and frog and sole dermis veins. The coronary vein, the independent superficial vein, the proximal branch of the caudal hoof vein, and the circumflex vein all drain the dermal lamella. The circumflex vein drains the toe and quarters.

The anastomoses of arteries and veins, which are blood vessels producing shunts, are another aspect of the horse foot's blood circulation. A meshwork of anastomosing arteriovenous veins is present at the base of each dermal papilla in the periople, coronary band, frog, sole, and terminal papillae. Because these anastomoses may drain nearly 50% of the total limb blood flow, they might be involved in ischemia owing to blood flow diversion. This may explain the link between laminitis and ischemia.

6.4.3. Bone Physiology

Bone is a complex, dynamic bone that may develop ontogenically, heal after injury, and respond adaptively to a range of external and endogenous stimuli. Through its unique physiological and biochemical qualities, osseous tissue allows bones to fulfill a variety of roles inside the animal's body. It is composed of a mineralized organic matrix in which the cells responsible for its production and renewal are lodged.

The organic matrix, or osteoid, is composed mostly of type-I collagen (about 95% type-I), which gives the bone its tensile strength, as well as trace quantities of other collagens and non-collagenous proteins whose primary function is to allow the matrix to mineralize. The main mineral salt present in osseous tissue is hydroxyapatite [Ca10(PO4)6(OH)2], which when attached to matrix proteins makes the tissue resistant to compressive stresses.

The bone is either cortical or trabecular in structure. Cortical, or compact, bone comprises the dense outer parts of the bone and amounts to 80 percent of total skeletal mass in humans. Cortical osteons, also known as Haversian systems, are tubular structures composed of a central channel (Haversian canal) through which a nerve and blood supply are provided, surrounded by coaxial lamellae of mineralized bone matrix containing several voids inhabited by cells in the form of lacunae and canaliculi.

Trabeculae (from the Latin trabs, meaning "beam") make up the remaining 20% of osseous tissue. which offer structural support to the surrounding cortex in a buttress-like fashion Trabecular osteons, or packets, have a lamellar architecture like that of the cortex, but are smaller and semi-lunar in form. This hierarchical design, a typical (though not universal) feature in biological materials, provides bones with the physical strength they require to perform their structural support, organ protection, and supplying leverage to muscles and tendons, aiding movement.

Three cell types are involved in bone remodeling and renewal: osteoblasts, osteoclasts, and osteocytes. Osteoblasts are mononucleated cells that are produced by the differentiation of mesenchymal stem cells and are in charge of the synthesis of osteoid and its subsequent mineralization.

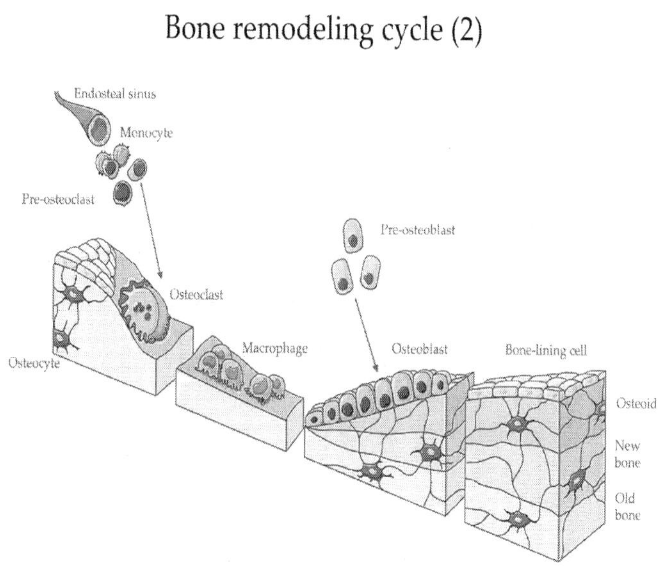

Figure 6.4. Bone Remodeling Cycle.

Source: Image by Wikimedia Commons

They are found primarily in the endosteum—the vascularised cellular lining of the internal proportions of the bone, covering the walls of the Haversian canals, trabeculae, and medullary cavity where it separates the bone matrix from the marrow—as well as the cambial layer of the periosteum—a vascularised and innervated structure consisting of an outer fibrous layer composed of fibroblasts, collagen. Osteoclasts are multinucleated macrophagic cells that are produced from phagocytes in the hematopoietic bone marrow and perform bone resorption in specific regions of the bone surface to which they attach when activated.

Osteocytes are the terminal development of osteoblasts that have been entrapped inside the bone matrix that they have synthesized and contain many cytoplasmic processes that cross the canaliculi of the osteons. They function as mechanoreceptors, interacting with osteogenic/osteolytic cells via gap junctions at the ends of their cytoplasmic processes, and regulate the bone synthesis/resorption cycle.

The synthesis/resorption cycle is involved in pH balance and mineral homeostasis, in addition to the physical necessity for the bone to be able to remodel for general maintenance, repair, and increasing structural strength in response to stimuli.

While the pH of the blood and extracellular fluid is normally kept within limited constraints by the elimination and excretion of protons by the kidneys and lungs, a variety of physiologically unfavorable situations (e.g., renal disease or strenuous exercise) can result in acidosis. Because of hydroxyapatite's basicity, bone acts as an emergency repository for the base, buffering acidity with the products of osteoclastic resorption.

This method is possible by osteoclast activation at low pH, a uniqueness in general cellular terms, and the synergistic inhibition of osteoblasts. Similarly, bone works as a storage site for calcium and phosphorus, making them available for mineral homeostasis maintenance. Calcium and phosphorus are essential for a variety of biological functions, and their homeostasis is regulated by the parathyroid glands, thyroid gland, and kidney, which affect intestinal absorption, renal reabsorption, and bone synthesis/ resorption mechanisms via the intermediary of parathyroid hormone, calcitriol (1,25-dihydroxyvitamin D, a hormone derived from vitamin D), and calcitonin, respectively. In reaction to a decrease in circulating calcium ions, the parathyroid glands, which contain Ca2+-sensing receptors, release PTH.

In the kidney, it decreases phosphate and increases calcium reabsorption, while in the bone, it stimulates osteocytic and osteoclastic activity.

Bone also serves as a reservoir for growth factors and cytokines, which are produced during bone resorption and can have a local or systemic effect. Along with their effects on other tissue cell types, such as endothelial cells of the vasculature, these growth factors and cytokines play important roles in bone formation and resorption, including insulin-like growth factors, transforming growth factors, and bone morphogenic proteins as growth factors promoting osteogenesis; epidermal growth factor, granulocyte-macrophage colony-stimulating factor, macrophage-colony stimulating factor, and tumor necrosis factor. As osteolysis-stimulating cytokines, prostaglandins, and leukotrienes; and interleukins, which may directly or indirectly promote bone formation or resorption depending on the interleukin family in question.

Along with a robust vascular and nerve present in the periosteum, many neurovascular bundles penetrate the bone through nutritional foramina, descend and climb the canals of Haversian systems, and reach the medullary cavity through Volkmann's Canals. The two systems are interrelated in that the bone relies on the vasculature for oxygen and nutrients, while hematopoiesis occurs in the bone marrow, which is controlled by osteoblasts.

6.4.4. Morphological Changes and Pathologies in the Foot

The outward morphology of the hoof capsule is related to the function and form of the hoof's interior segments. Dyson and colleagues noted that, despite variances in distal phalanx orientation amongst horses, which are mostly due to alterations in the direction of the dished solar border, the morphology of the distal phalanx is unaffected by the exterior characteristics of the hoof capsule. It is important to note that trimming and shoeing can change the form of the hoof.

The effect of trimming/shoeing on hoof capsule shape was described, and the researchers established that the creation of the hoof wall is physically tied to the loading of the lower limb, protecting its proper balance on the ground. As a result, the geometrical tendency of the foot components impacts the internal structures' capacity to adapt to loads during the bearing part of the stride cycle.

The distal phalanx is joined within the hoof capsule by the suspensory apparatus, which binds the entire parietal surface of the distal phalanx to the internal hoof wall's lamellar structures. Preliminary research on the horse lamellar connection discovered that this attachment offers the mechanism for weight transmission between the distal phalanx and the hoof wall's epidermal laminae.

This link, or attachment, plays an important function in the biomechanics of optimal foot performance and, if broken, can contribute to foot lameness. Indeed, the loss of the link between the epidermal laminae and the dermal lamellae's underlying basement membrane would impair the distal phalanx's suspensory apparatus. Changes in the suspensory tissue's basement membrane, unsurprisingly, have been proposed to signify the earliest stage of laminar collapse. Other studies have found that lesions in the basement membrane emerge before clinical indications of foot lameness.

The extensor process of the distal phalanx puts pressure, first on the sole at the palmar border of the distal phalanx and then on the coronet or top portion of the lamellae, through distal phalanx dislocation and rotation. These deflections reduce blood flow into the hoof wall's basal layers and can slow the development rate of the hoof capsule, changing its form over time and causing osteolysis of bone trabeculae in chronic phases. This chronic illness can cause a variety of hoof morphologies, including shorn heels, crushed heels, club foot, long-dished toe, and high-heel foot.

The lamellar wedge, which develops with laminitis and can result in morphological displacement of the distal phalanx inside the hoof capsule, is a direct outcome of the failure of the distal phalanx's suspensory mechanism. The molecular processes involved in the lamellar wedge state, on the other hand, remain mostly unknown. The lamellar wedge shows in chronically laminitic horses as an aberrant horny mass produced between the inner hoof wall and the epidermal lamellae and is associated with a little rotation of the distal phalanx.

Due to the uneven development of the proximal hoof wall part, the separation of the distal phalanx inside the capsular wall might cause the sole to become convex rather than concave. Because the structural and physical appearance of this aberrant keratinized material is similar to that of white line tissue, it is thought to be an ectopic white line. As a result, it was assumed that a substantial amount of the ectopic white line may eventually prevent the straight and regular formation of the hoof capsule.

6.5. MYOCARDIAL METABOLISM

Modifications in myocardial metabolism are connected with myocardial dystrophy and result in heart chamber dilatation, reduced contractility, organ perfusion, and symptoms. In veterinary medicine, heart failure therapy now covers neurohormonal, circulatory, and contractile elements of this diseased condition. Unfortunately, the energy-supply component is not included in recent guidelines. Most commonly used drugs alter contractile capacity via controlling myocardial filament sensitivity to various ions, but do not influence cardiomyocytes' ability to create adequate energy for this task.

One of the most energy-dependent structures is the myocardium. It requires around 6 kilos of ATP every day. It features an innovative mechanism that produces enough ATP to maintain an efficient energy supply. There are two methods for the body to meet this demand: production and accumulation. Accumulation is unsuitable for the heart because of its unique anatomy—the majority of the cytoplasms are made up of myofibrils.

According to this fact, we find low ATP concentrations and a high number of ATP-hydrolases in the adult heart. In normal myocardium, total ATP volume resynthesis takes only 10 seconds. Most of the energy resources (70%) are used for contraction, with the remainder employed for ion pump function (K, Na, Ca pumps ATPases).

This system is highly coordinated, which contributes to the regular flux of energy substrates and ions. The heart eats around 20 g of carbs, 30 g of free fatty acids (FFA), and triglycerides on average (TG). In 35 L of oxygen, these substrates are oxidized to create ATP from ADP.

FFA oxidative phosphorylation produces around 60% of all ATP generated, whereas glucose, lactate, and other carbohydrates oxidation produce approximately 30% of all macroergic chemicals. Ketone bodies and amino acids can also be used as energy supplements.

In some circumstances, glucose consumption might be the primary source of energy (a high-carbohydrate diet). As a result, alterations in myocardial metabolic characteristics during heart failure may give critical information for the early diagnosis and treatment of myocardial disorders.

Cardiac failure syndrome is a complication of major heart disease and is linked with compensatory mechanism malfunction, the establishment of abnormal connections between components of neurohumoral regulatory systems, and their activation. Decompensation is a condition characterized by decreased energy output and inhibited cardiac metabolism.

Systolic dysfunction, for example, causes sympathoadrenal system hyperactivity, which is related to a higher heart rate. Catecholamines stimulate beta-adrenergic receptors, increasing myocardial oxygen use due to increased FFA usage to provide sufficient energy. This scenario results in increased ADP volume and a negative inotropic impact, which is poorly tolerated during heart failure and progresses geometrically with prolonged sympathetic tonus.

6.5.1. Metabolism in the Adult Healthy Heart

Carbohydrates and free fatty acids are the primary substrates for ATP generation. Long-chained FFA, glucose, glycogen, lactate, pyruvate, ketone bodies (acetoacetate, beta-hydroxybutyrate), and amino acids, in particular (leucine, valine, and isoleucine) are substrates. These chemicals are broken down into intermediates that enter the Krebs cycle as acetyl-coenzyme A (A-CoA) or other metabolic equivalents.

The proton is produced during substrate usage. This proton generates an energetic gradient between the mitochondrial membranes, stimulating the oxidative chain to generate chemical energy and phosphorylate ADP to ATP.

Such a range of substrates for the creation of common energy sources predisposes to numerous concepts:

(1) myocardial metabolism is very adaptable to organism conditions and substrate environment, and can switch between main energy resources; unfortunately, in heart failure, this flexibility is mostly lost.

(2) myocardial metabolism is a self-regulated mechanism; all intermediates of the tricarboxylic acid cycle are mediators, controlling the main metabolic path and intensity of energy production (Randle cycle).

(3) metabolites can be used as components for cell structure synthesis, whereas cellular structures might be employed as an energy substrate.

(4) Metabolic dysfunction and metabolite accumulation can damage cellular proteins and alter the shape and function of contractile filaments.

(5) myocardial metabolism is not "intracellular chemistry"; it is a functional system with specific structure and mediator mechanisms that assesses cardiomyocyte adaptation to environmental variations.

The efficiency of myocardial metabolism is strongly reliant on the pathways and substrates used for ATP generation. A Kyoto Encyclopedia of Genes and Genomes (KEGG) database exists, which is a collaboratively constructed map of known molecular interactions and feedback loops of energy metabolism in the heart. This created genome map aids in understanding different pathways of energy production in the myocardium and limiting its activity. However, we should look for common characteristics of cardiac metabolism.

Under aerobic circumstances, mitochondrial oxidative phosphorylation produces the majority of ATP (approximately 90%); the remaining macroergic chemicals are created by anaerobic use. Mitochondrial oxidative phosphorylation generates energy by FADH and NADH dehydration, which is obtained through FFA beta-oxidation and, to a lesser extent, other sources. Stanley et al. schematic's system of metabolic interactions depicts the key aspects of energy generation cycles.

FFA transport in the cardiomyocyte occurs via two mechanisms: passive diffusion and specialized protein transporters. Long-chained FFA is distributed throughout the cell, metabolized in acyl-CoA, and delivered to mitochondrial membrane special proteins to interact with acetyl-CoA synthase. FATP1, FATP6, and CD36 support active transport, which is stimulated by muscle contraction or insulin (Ins) action. These proteins carry FFA across the membrane and link it with CoA, which is then carried to the lipid beta-oxidation cycle via carnitine-associated translocators.

Furthermore, cytosolic carnitine palmitoyltransferase-1 (CPT-1) links acyl-CoA with carnitine to create long-chained acylcarnitine. This molecule is carried across the inner mitochondrial membrane by acylcarnitine translocase and used in FFA beta-oxidation cycles with acetyl-CoA synthesis. The Krebs cycle then converts acetyl-CoA to ATP, H2O, and CO2. In the tricarboxylic acid cycle, for example, palmitate is oxidized with 23 moles of O2 to create 105 moles of ATP.

Nonetheless, due to their high oxygen consumption, FFA is an ineffective energy source when compared to glucose. A portion of the FFA delivered is esterified and collected in the cytoplasm as lipid droplets (triacylglycerol-TAG). In physiological settings, TAG-produced ATP accounts for around 10% of total acquired ATP. TAG is also a significant component of FFA oxidation in circumstances where TAG-hydrolase is inhibited, lipid beta-oxidation is significantly diminished, resulting in large lipid droplet build up in cardiomyocytes.

The Krebs cycle will then be activated. This rotor begins with acetyl-CoA, which is obtained from FFA beta-oxidation or pyruvate decarboxylation. NADH and FADH2 transporters produced are equal to an electron chain, stimulating ATP resynthesis in oxidative phosphorylation.

Generation metabolic pathways are governed by directing components (enzymes) and feedback connections (substrate-final product). The mitochondria can withstand high-energy demand conditions, boosting oxygen consumption by about 85 percent over the baseline. This skill is critical since it requires just 25% of the oxidative capacity most of the time.

As a result, activation/inhibition of enzymatic systems can govern ATP generation and, through feedback, adjust energy substrates in instances of excessive metabolite collection or regulatory abnormalities. This type of metabolic flexibility is extremely beneficial in cardiac illnesses caused or influenced by a lack of energy resources.

Furthermore, under normal conditions, the myocardium uses lactate, which is metabolized to pyruvate by lactate dehydrogenase and enters the Krebs cycle. When a metabolic problem occurs, the myocardium begins to secrete lactate into the circulation. This method emerges when there is a lack of oxygen and energy must be created by anaerobic glycolysis (ischemia, terminal stages of cardiomyopathies).

The monocarboxylate transporter is the primary transporter mediating lactate excretion and intake (MCT). This family has four subclasses, however, only one form of MCT-1 is found in the myocardium. They also play a role in ketone body transfer.

Glycolysis is another coexisting energy generation route. The first step in glycolysis is the transfer of glucose across the cell membrane by a particular transporter (GLUT). Glucose is converted in the cytoplasm to pyruvate, which is then transferred to the mitochondrial matrix by pyruvate dehydrogenase (PDH). Pyruvate is converted to acetyl-CoA and enters the Krebs cycle.

6.5.2. Regulation of Carbohydrates and FFA Oxidation

FFA use is the primary regulator of carbohydrate oxidation. Increased FFA intake promotes intermediate buildup, which inhibits PDH. Simultaneously, reduced FFA intake enhances glycolysis and lactate oxidation due to a lack of citrate, NADH, and acetyl-CoA in the mitochondrial matrix. The latter is frequently observed in cardiomyopathies and during ischemia.

Furthermore, cytosolic carnitine palmitoyltransferase-1 (CPT-1) links acyl-CoA with carnitine to create long-chained acylcarnitine. This substance is small molecule proteins—energy homeostasis regulators—that have been shown to have an important function in modern research. Of course, there are other substances and variables that influence energy metabolism; one increases appetite (ghrelin, galanin, neuropeptide Y), while another is anorexigenic (leptin, nesfatin-1).

Insulin (Ins) was the first discovered molecule that governs energy homeostasis; its activity was originally observed as neurogenic appetite suppression. Later, leptin was discovered—a hormone generated by adipose tissue that elucidates the general status of adipose tissue. Ghrelin and nesfatin-1 were discovered to have antagonistic effects on adipose tissue to leptin. Over the next few decades, there was a lot of study on lipid homeostasis and appetite-controlling peptides. Many molecules were discovered, with the most significant being preptin, irisin, and adropin.

Insulin (Ins) is a hormone with a wide range of physiological effects, but we will just examine three of them here: heart pump performance, Ca+2 ion circulation, and as a mediator between cell communication. The fundamental mechanism of energy production of membrane-associated ATPases and ion pumps is ins-induced glucose transport.

Ins controls cytoplasmic concentration and equilibration of Ca+2 indirectly by controlling pump activity; it facilitates cascades of events to drive Ca consumption or excretion. Ins regulates NO generation and tissue perfusion and is implicated in endothelial function (including coronary vessels). And, of course, it influences cardiomyocyte contractile capacity owing to energy metabolism regulation.

Ins' capacity to influence the availability of energy substrates (effects on liver and adipose tissue) and tissue perfusion should be acknowledged as an indirect impact. Ins inhibits TG hydrolysis in adipose tissue (depressing lipomobilising hormones), lowering circulating FFA levels. Furthermore, reactive ins secretion enhances tissue perfusion owing to smooth muscle relaxation in blood vessels. This impact is particularly important during exertion, hypertension, and acute and chronic heart failure.

Direct Ins activity controls important enzymes and transporters (6-phosphofructokinase 1 and 2, glycogen phosphorylase and synthase, PDH, hormone-dependent lipase, acetyl-CoA carboxylase). Randle's cycle elucidates interactions between major metabolic substrates (glucose and FFA). Ins regulates transmembrane glucose transport by GLUT 1 and GLUT 4. (both transporters are Ins-determined, but GLUT 1 is less dependent).

GLUT 4 is abundant in myocardium tissue, which aids in maintaining myocardial energy flexibility during exercise and heart failure. Ins has a number of effects on glycogen accumulation: reducing glucose utilization (predominance of FFA oxidation results in PDH blockage, glucose intermediates converted to glycogen); HX2 capacity in overloaded glucose transport (Ins-dependent GLUT 4 exocytosis); glycogen utilization in glucose deficiency. It should be noted that glycogen is metabolized more aggressively than glucose due to its intracellular position and higher ATP generation. Furthermore, Ins activates glycogen synthase both directly and indirectly via G6P increase. The interactions between Ins and PDH are unclear. The impacts of FFA oxidation suppression (lower acetyl-CoA concentration in mitochondria), influence on PDH phosphatase, NAD/NADH ratio, and Ca+2 concentration should all be considered.

In general, Ins regulates glycolysis indirectly via metabolite and substrate availability and directly via enzymatic mechanisms (mentioned above). Ins' influence on FFA oxidation is strongly related to its effects on glycolysis, which are briefly discussed above. Ins, by the way, inhibits CPT-1 action owing to malonyl-CoA concentration. It is explained by the fact that acetyl-CoA carboxylase, which is directly controlled by Ins, produces malonyl-CoA.

Based on metabolic influence, the mediator effect of ins between cells is defined by its effects on PDH, HX2, phosphofructokinase, glycogen synthase, acetyl-CoA carboxylase, hormone-dependent lipase, PDH kinase, MAP kinase, and lactate intercellular shuttle.

In recent studies on dogs with chronic degenerative valve disease, the benefits of leptin were discovered. The investigation discovered an increase in circulating leptin and leptin microRNA in this condition. Because the dogs were not obese, it was discovered that leptin alterations are associated with heart failure syndrome. Furthermore, a link was discovered between leptin levels and the severity of heart failure.

Preptin is a hormone that regulates glucose metabolism and is a member of the insulin family (as insulin, insulin-like growth factor-1, proinsulin-like factor-2, and relaxin-2). It was discovered in tests that it is released alongside insulin and increases glucose utilization in insulin-like ways. Preptin expression was found to be strongly linked to insulin resistance. In general, this hormone influences hepatic glycogenesis and bone density (osteoclast proliferation), as well as insulin sensitivity and adaptability to energy substrates.

Adropin is a recently discovered hormone that regulates lipid metabolism. Depending on the food, adropin affects energy metabolism (significantly raised on a high-fat diet). Adropin treatment systemically reduces the severity of hepatosteatosis and hyperinsulinemia (moderating carbohydrate-FFA metabolism in peripheral tissues).

A link between heart failure severity and circulating adropin concentration was discovered in studies (high severity of heart failure-high adropin level). In diabetes mellitus, the amount of circulating adropin is also reduced in insulin resistance and is associated with an increased risk of atherosclerosis. Low levels of adropin were linked to endothelial dysfunction and an increased risk of heart X syndrome. Adropin inhibits PDH kinase 4, promoting normal pyruvate use in the Krebs cycle, and lowers CPT-1 activity and CD36 transporter traffic, reducing FFA transport in cardiomyocytes.

Adropin's primary roles include regulating NO availability, reducing lipogenic gene expression, lowering dyslipidemia and hepatic steatosis, changing insulin resistance and glucose tolerance, and modulating metabolic homeostasis.

Irisin is a hormone that regulates the transformation of white to brown adipose tissue. White adipose tissue lacks mitochondria but is high in TG and FFA and generates leptin, ghrelin, and nesfatin-1. While brown adipose tissue is densely packed with mitochondria and lipid droplets. This cell contains a high concentration of uncoupling protein-1. This protein enhances heat generation by uncoupling ATP creation from FFA oxidation rather than ADP phosphorylation.

Experiments revealed that large levels of circulating irisin are prevalent in instances of obesity, a condition known as irisin resistance (insulin resistance-like). During the activity, skeletal muscles generate the majority of irisin. This hormone's major activities are to decrease white adipose tissue, manage temperature homeostasis, and increase in glucose tolerance, obesity reduction, and insulin resistance modulation.

6.5.3. Energy Substrates and Contractility

Energy status changes (ATP/AMP ratio fluctuation), increased intercellular Ca^{+2} build-up, stretch, GLUT 4 exocytosis, glucose and FFA intake, and other factors all contribute to muscle contraction.

Many studies have shown that cardiac contractility is highly effective in settings of intense glucose utilization; nevertheless, increased FFA intake by 26% did not result in an equal rise in contractility, but rather in an increase in oxygen demand. Targeting FFA oxidation processes and FFA binding to inaccessible molecules reduces oxygen consumption while increasing the mechanical power of rat cardiac contraction.

The combination of insulin and glucose induces a 39% reduction in the heart's oxygen consumption. These consequences are unclear because, theoretically, palmitate or oleate usage requires fewer molecules of O2 to make one molecule of ATP than glucose or lactate. Interactions between long-chained FFA and Ca+2 channels are one probable reason (increases ATP demand for a pump ATP-ase.

Recent research has shown that elevated FFA and TG concentrations in the cytoplasm can cause lipotoxicity in the heart, resulting in neutral lipid and ceramide buildup, cell death, and reduced contractility. In tests, Zhou demonstrated that large levels of TG and ceramides were accumulated in diabetic rats, inducing DCM-phenotype alterations, reduced contractility, and significant levels of cardiomyocyte apoptosis.

Nonetheless, in the case of troglitazone, the manifestation of the FFA block was greatly reduced. Lipid-induced myocardial remodeling is mostly unclear at this time; however, it may be related to cell death, decreased contractility due to extensive FFA use, and considerably reduced glycolysis. Regardless of origin, the development of heart failure syndrome is always coupled with an energy deficit. Individual cardiomyocytes are under a heightened burden because of the high demand for macroergic substrates during this condition, while their production is severely reduced.

Because of decreasing levels of creatine phosphate and ATP, this situation is referred described as an engine out of fuel. Compensatory and pathogenic cardiomyocyte enlargement is related to a drop in the creatine phosphate/ATP ratio, as well as a subsequent decrease in ATP. The creatine phosphate/ATP ratio is a solid predictive marker for worsening heart failure.

6.5.4. Myocardial Metabolism in Heart Failure

Heart failure causes myocardial metabolism to become less flexible. HF has a propensity to convert FFA use as the primary energy source to glucose oxidation at some phases. This stage is distinguished by decreased FFA intake, diminished FFA oxidation enzymes, and mitochondrial oxidation indicators. This change is generally seen right away.

It was discovered in tests that metabolic alterations in rat myocardial occur in the second week after artificial aortic constriction, but reduced contractility appears only on the 20th week after bandage. According to some experts, glycolysis predomination is a sign of terminal cardiac metabolic failure. These changes are related to adaptability since glycolysis requires 13% less oxygen to create the same quantity of ATP as FFA oxidation.

The switch to glycolysis increases glucose intake and GLUT 1 expression. At the same time, glucose oxidation is changed, resulting in the uncoupling of glycolysis and glucose oxidation. The combination of reduced FFA consumption and glucose oxidation results in a reduction in mitochondrial oxidative potential.

Pyruvate is not delivered to the mitochondria during glycolysis and glucose oxidation uncoupling due to PDH suppression by PDK, but is converted to lactate by LDH. This causes cellular acidosis, and by the way, anaerobic glucose consumption yields just two molecules of ATP (whereas aerobic yields 32). Cardiomyocyte enlargement, energy metabolism depression, ionic pump malfunction, Ca+2 build-up, reduced contractility, apoptosis, and fibrosis are all promoted by the described modifications.

It should be highlighted that this pattern of myocardial dysfunction development is consistent across all cardiomyocytes; even in situations of pulmonary hypertension and compensatory hypertrophy of the right heart, metabolic changes will be identical to those seen in left heart failure.

According to existing evidence, heart failure enhances myocardial tissue insulin resistance, which is largely attributable to neurohormonal remodeling, and is an independent predictor of sudden cardiac death in people. Insulin resistance impairs glucose consumption and ATP generation. According to certain research, TG buildup in muscles (discovered using the 1H NMR approach) enhances insulin resistance.

Randle's cycle explains the relationship between TG buildup and insulin resistance: high FFA intracellular accumulation promotes increased acetyl-CoA/CoA and NADH/NAD ratios, which block PDH and lead to citrate accumulation and phosphofructokinase inhibition. G6P buildup in the cell inhibits HX2, boosting intracellular glucose accumulation while lowering intracellular glucose transport.

High circulating insulin concentrations can also be related to insulin resistance. Adrenergic hyperactivity, which occurs in conjunction with heart failure, causes accelerated glucose mobilization, hormone circulation, and insulin production, as well as lipomobilisation owing to catecholamines (noradrenaline). Insulin promotes GLUT 4 and CD36 exocytosis and, in the early phases, aids in the production of ATP via glycolysis and oxidative phosphorylation.

However, the activation mechanism of insulin receptors varies. Insulin receptors have two points of contact with insulin. One has a high affinity for hormones and encourages a rapid response to insulin stimulation; the other is a "slow" one that is activated in circumstances of high insulin concentration and partially blocks the "fast" section of the receptor owing to geometrical conformation.

Insulin resistance is characterized by the blockage of all "fast" receptors, increased insulin concentration, and hormone level on the "slow" locus of the insulin receptor. Furthermore, a high circulating FFA content reduces insulin-stimulated GLUT 4 translocation. This is explained by the suppression of IR-1 Pi 3 kinase, whose phosphorylation is reduced by TG and phospholipid (FFA-acetyl-CoA, diacylglycerol, ceramides) buildup in the cytoplasm. This process is also aided by enhanced GLUT 1 expression. Increased GLUT 1 glucose flow supports lower GLUT 4 exocytosis and higher GLUT 4 tissue concentration.

Pathological cardiomyocyte hypertrophy and systolic dysfunction arise as GLUT 4 function is reduced. Another reason is the use of pyruvate in anaplerotic processes, which results in less acetyl-CoA generation for Kreb's cycle, uncoupling of glycolysis and oxidative phosphorylation, as well as activation of PDK 4 (promotes suppression of insulin-stimulated glycolysis).

It should also be highlighted that HX2 activity is reduced in diabetes and insulin resistance. Insulin was discovered to be an HX2 gene expression and protein resynthesis regulator in cell culture research. As a result, the severity of insulin resistance acts as a suppressor of HX2 activity, resulting in G6P accumulation and protein glycosylation in the cytoplasm.

It should be acknowledged that decreasing HX2 microRNA levels are linked to GLUT 4 gene expression and protein depletion. Insulin sensitization—by thiazolidinediones (pioglitazone, troglitazone)—can modulate these interactions between insulin, HX2, and GLUT).

Heart failure is frequently accompanied by all energy-producing enzyme malfunction. Creatine kinase (CK) activity has been shown to be significantly reduced. This enzyme controls the transfer of ATP and creatine. CK is a dimer composed of two components M and B, with three isoforms: MM, BB, MB, and mitochondrial-CK. MM-CK is tightly associated with SR and Ca+2-ATPase, creating energy for Ca+2 circulation.

Mitochondrial-CK is found on the mitochondrial inner membrane and collaborates with the ADP-ATP translocator. Translocator transports produced ATP to mitochondrial-CK and then to creatine phosphate or ADP. This compartment distribution controls the local ATP/ADP ratio well, promoting mitochondrial ATP generation (lower ratio) or increasing enzyme activity. However, in cases of cardiomyopathy, the typical compartment system has been revised. Decompartmentalisation causes the mitochondria—mitochondrial-CK-ATP and phosphocreatine connections to decouple.

6.6. VETERINARIAN'S ROLE IN CONSERVATION MEDICINE AND ANIMAL WELFARE

Human actions have increased the number of conservation issues and developing illnesses, putting a strain on the long-term survival of vulnerable free-ranging and captive animal species while also endangering ecosystems and public health. Veterinarians not only have a comprehensive education in comparative medicine (rather than a single-species specialization), but they are also highly skilled in identifying, diagnosing, and comprehending the impact of disease on public health as well as people, groups, and entire ecosystems.

Their knowledge and experience make them vital essential actors in the design, implementation, and successful assistance of both *in-situ* and *ex-situ* conservation initiatives. Major aims have now taken precedence in parks and zoological gardens: the conservation of global wildlife and flora, as well as the preservation of animal welfare. Animal welfare may now be objectively analyzed to determine an individual's quality of life, with behavioral assessment and behavioral enrichment being key techniques.

The present rate of extinction is extraordinary (approximately 1000 times higher than previously predicted) and is likely to be underestimated if biodiversity data are influenced by a deficit in information on species taxonomy, range, and status. Human actions have increased the number of conservation issues and developing illnesses, putting a strain on the long-term survival of vulnerable free-ranging and captive animal species while also endangering ecosystems and public health.

Biodiversity conservation now necessitates a broader approach capable of connecting interdisciplinary bridges between the diversity of social, economic, political, and biological variables that influence health issues, in order to understand their complex interaction and thus find new solutions while keeping human and animal well-being in mind while preserving ecosystems, referred to as the "OneHealth" approach.

Zoological and wildlife medicines are recognized scientific areas of the veterinary profession that evolved in the 1990s in response to growing concerns about wildlife conservation and now fully integrate the transdisciplinary frame that characterizes conservation Biology.

Wildlife veterinarians and the zoological community play an important role in the future of biodiversity conservation because, in addition to saving species and understanding the effects of animal diseases on human communities, they also protect functional and integrated ecosystems by applying their skills and scientific understanding in the growing area of Preservation Medicine.

Contemporary zoos and aquariums have a responsibility to the animals in their care throughout their whole life cycle. Animal behavior may be radically altered in captivity. By enclosing animals in a cage or enclosure, we diminish the complexity of their environment, severely limiting their natural control over it and limiting the range of behaviors they may display.

Where animals have little options, we are the ones that design practically every element of their lives (e.g., feeding schedules, what to eat, where to sleep, who to live or to reproduce with). Environmental sensory deprivation and lack of physical variation can lead to hostility, boredom, anxiety, frustration, and, eventually, physical and physiological sickness.

Moreover, the maintenance of basic biological behaviors is critical to the survival of individuals intended for release and reintroduction into the wild, and hence to the success of conservation projects. As a result, captive facilities have an ethical and legal duty to ensure the holistic welfare of all animals under their care. They should collaborate in order to attain high standards of animal welfare (AW), meet the diverse requirements of animals, and reduce the occurrence of negative states while boosting good ones.

This includes:

(1) providing acceptable, safe, and naturalistic habitats.

(2) giving sufficient food.

(3) providing necessary veterinarian care.

(4) providing appropriate social contact and

(5) offering environmental enrichment.

Overall assessment of AW is not easy and should be done in a scientific and objective manner, avoiding anthropomorphism and taking no account of ethical issues about the activities or situations being compared in its evaluation.

6.7. VETERINARIANS' ROLE IN ANIMAL WELFARE AND BEHAVIORAL ASSESSMENT

The veterinary profession operates under an ethical code of professional behavior that requires adherence to integrated animal welfare principles as well as the individual duty of ensuring the reasonable use of the "Five Freedoms." All animals must be treated with dignity, respect, and compassion, as well as with due regard for their species-specific biology and behavior.

Despite tremendous breakthroughs in the field of animal welfare in recent years, the majority of study has focused on farm and domesticated animals. Many poorly understood species, as well as individuals with diverse life experiences and temperaments, remain in zoological institutions. Animal welfare is a broad, multidisciplinary subject with several meanings.

The Brambell Report on the welfare of intensively farmed animals, issued by the British government in 1965 and later revised by the Farm Animal Welfare Council (FAWC) in 1979, resulted in the declaration of five formalized and rightful freedoms that would form a logical and comprehensive system for animal welfare analysis. According to the Farm Animal Wellbeing Council, an animal's welfare involves both its physical health and emotional condition, and "every animal held by man must at the very least be safeguarded from unnecessary suffering."

Although "perfect accomplishment of all five freedoms is unattainable," according to Webster, they nonetheless constitute an "effort to make the most of a complicated and challenging circumstance. "A few of these liberties, such as freedom from fear and anxiety or pain, are anthropocentric constructions. In suitable settings, fear and pain are normal and necessary defensive mechanisms that may have adaptive and fitness value".

To summaries, the freedoms establish "ideal states rather than norms for acceptable wellbeing" and are best considered as valuable and practical concepts that give the core philosophy to avoid pain and achieve a condition of good welfare and evaluation of any husbandry system.

Figure 6.5. The Five Freedoms of Animal Welfare.

Source: Image by Flickr

Animal wellbeing is a scientific discipline of applied animal behavior that may be quantified through behavioral assessments. Understanding behavioral diversity, its purpose, and its relationship to the animal's perception of its surroundings is crucial because it may assist to prevent what is deemed an aberrant demeanor or enhancing settings that are suited for typical repertoires.

Skinner defines behavior as "part of an organism's overall activity" and "that element of the functional organism that is involved in acting on or having trade with the outside environment." As a specialized reaction, behavior is the first line of defense against environmental demanding stimuli and can provide a rough sketch of an animal's coping success against external stresses. It is crucial in addressing Dawkin's questions:

(1) "Are the animals healthy?"

(2) "Do they have what they want?"

Since it covers the animals' own decision-making process and reflects a phenotypic representation of emotions, and it may also be employed in the clinical assessment of the animal's health state (e.g., assessment of pain, nutritional requirements and hormonal conditions).

Behavior assessment via applied behavior analysis and behavioral monitoring studies is a technique with various benefits that promote optimum animal care, making it critical to enhance animal welfare and satisfy conservation goals. Such studies have focused on scientific data collecting by direct and objective observation of observable behavior as well as the settings in which it occurs, and they are concerned with the functional links between environment and behavioral manifestations.

Systematic observations and record-keeping have numerous advantages as a management tool in zoos and other related facilities: they are a non-invasive and, in most cases, non-intrusive technique that allows documentation of normal behavior patterns and identification of any changes in regular activity, as well as the establishment of a database of background knowledge on individuals on a constant basis.

6.7.1. Behavioral Enrichment

Enriched history begins in the 1920s with Robert Mearns Yerkes, a psychobiologist most known for the work in intelligence testing of both humans and monkeys, as well as the articles on the necessity of enrichment for captive gorillas and chimps. Heini Hediger, renowned as the "Father of Zoo Biology," was a visionary in the field of proxemics in animal behavior.

It spoke on human responsibility in creating constructive settings for wild animals in zoos in the 1950s and underlined the necessity of understanding animal territorial surrounds, having a particular impact on the building and layout of naturalistic enclosures. This book, states, "Anyone who sets out to design homes for animals should be perfectly aware that the cube is the most non-biological and hence most improper of all spatial configurations."

Behavioral enrichment, also known as environmental enrichment, is today a scientifically proven component of animal husbandry. It should be thoroughly integrated into animals' daily routines as a technique for improving their quality of life and discouraging undesired habits that arise as "artefacts of confinement" (e.g., stereotypes).

Officially, enrichment is described as "a dynamic process for improving animal surroundings in the context of the animals' behavioral biology and natural history." Environmental alterations are undertaken with the purpose of expanding the animals' behavioral options and eliciting species-appropriate behaviors, hence improving animal wellbeing.

"The goals of behavioral enrichment can be met by creating productive environments that encourage each animal to express the species' natural mental activities and behavioral repertoire, as well as by adding stimuli that add complexity and novelty to its routine, as well as opportunities that allow it to regain the sense of control it should have over its environment".

Behavioral programs play a role in combating inactivity and obesity, reducing/eliminating stereotyped and aberrant behavior (e.g., aggression, sexual frustration), and lowering stress levels, which can, on their own, encourage reproduction. The 'S.P.I.D.E.R.' system, established by Disney's Animal Kingdom, is a solid paradigm that serves as a significant instrument in the conception, implementation, and management of institutional training and enrichment programs. These programs offer all taxa suitable challenges, opportunities, and stimulation. S.P.I.D.E.R. is an abbreviation for the initial letter of each system component, as follows: Goal setting; planning; implementation; documentation; evaluation; and re-adjustment the last component of the system occurs throughout the whole process of developing an enrichment plan.

6.8. CONCLUSION

Evidence suggests that the earliest appearance of scientific anatomical study may have been in ancient Babylonia. The ancient Egyptian civilization is believed to be the origin of the anatomist. Hippocrates described the human brain as being in two halves divided by a thin vertical membrane, as it was in other animals. The most renowned Alexandrian physicians were Herophilus and Erasistratus. The appearance of anatomist Mondino dei Luzzi in the thirteenth century induced progression in the anatomical and physiological study based upon his work with human cadaver dissection.

William Harvey was the first to correctly describe the circulatory system. The Anatomy Act (1832) was enforced in order to regulate the acquisition of human cadaveric material. This regulation favored anatomy schools based in hospitals and as such, the professionalization of the science was observed. At the present time, anatomy and physiology are an unarguably fundamental part of medical education. Recent developments in technology have allowed digital anatomical models to be implemented into university curricula, allowing wider access to the study of anatomy. Recent developments in technology have allowed digital anatomical models to be implemented into university curricula, allowing wider access to the study of anatomy for the contemporary student. Towards the end of chapter, it also discussed about the veterianrian's role in conservartion medicine and animal welfare, and their behavioral assessements.

REFERENCES

1. Al-Agele, R., Paul, E., Kubale Dvojmoc, V., J. Sturrock, C., Rauch, C. and Sian Rutland, C., 2019. The Anatomy, Histology and Physiology of the Healthy and Lame Equine Hoof. Veterinary Anatomy and Physiology, [online] Available at: <https://www.intechopen.com/chapters/65570> [Accessed 22 June 2022].
2. Kubale, V., Cousins, E., Bailey, C., A.A. El-Gendy, S. and Sian Rutland, C., 2019. Introductory Chapter: Veterinary Anatomy and Physiology. Veterinary Anatomy and Physiology, [online] Available at: <https://www.intechopen.com/chapters/65366> [Accessed 22 June 2022].
3. Oleinikov, D., 2019. Myocardial Metabolism. Veterinary Anatomy and Physiology, [online] Available at: <https://www.intechopen.com/chapters/63726> [Accessed 22 June 2022].
4. Raquel Neves Fernandes, D. and de Lurdes Ribeiro Pinto, M., 2019. Veterinarian's Role in Conservation Medicine and Animal Welfare. Veterinary Anatomy and Physiology, [online] Available at: <https://www.intechopen.com/chapters/65450> [Accessed 22 June 2022].

CHAPTER 7

CURRENT TRENDS IN TEACHING AND LEARNING ANATOMY

CONTENTS

Learning anatomy is a key in the medical field, mainly for the people who are heading towards the medical sciences. The chapter provides historical context initially followed by some methods, models and tools of teaching. Evolving trends have also been discussed from a global perspective. Later, the basics of a VR system for anatomy training, its preparation and integration models are defined. Towards the end of the chapter, some challenges and opportunities in teaching anatomy have been described.

7.1. INTRODUCTION

The phrases "anatomy," "anatomy course," and/or "morphology classes" conjure up a slew of memories, ranging from happy ones like appreciation and nostalgia to bitter ones like resentment, frustration, and despair. In any event, one thing is clear: for some, anatomy lectures and the rest of the "basic" medical courses were a necessary evil, while for others, they were a passion and a way of life.

However, they were, are, and will continue to be a part of our medical education as well as that of other healthcare professionals, and they are an invaluable resource for several colleagues who regard anatomy as the backbone of their clinical practice or medical-surgical specialization, and/or an almost limitless source of advances in diagnostic imaging, surgical procedures, rehabilitation, and bodily restoration processes, to name a few.

Regardless of country, racial background, or medical school system, anatomy has always been a cornerstone in medical education. Medical students gain a first "impression" of the structure of the human body through learning gross anatomy, which is the foundation for understanding pathologic and clinical problems. Although the significance of teaching anatomy to undergraduate and graduate students is undeniable, there is now a heated discussion over how anatomy should be taught. In the past century, dissection and lectures were its sole pedagogy worldwide.

Recently, the amount of time allotted for anatomy instruction has been drastically reduced, to the point where some argue that it is no longer up to par. Traditional anatomy education, which consisted of topographical structural anatomy lectures and gross dissection classes, has been replaced by a variety of study modules, including problem-based learning, plastic models or computer-assisted learning, and curricula integration.

"Does the anatomical theatre have a future in medical school?" "What exactly is the issue with anatomic specimens?" The aim is to provide answers to both of these concerns as well as contribute to the discussion about the current state of anatomy instruction at the undergraduate and graduate levels.

Anatomy (health descriptions) is defined as "a branch of biology concerned with the structure of organisms" or "a science concerned with the structure of the human body." However, its definition and reach are far broader. Anatomy is a fundamental science that enables healthcare professionals in training, like undergraduates and graduates, to gain a completer and more comprehensive grasp of what it means to study the human body in order to situate it in the dynamic situations of health, sickness, and disability.

The technical and anatomical terminology required for pharmacology, pathology, physiology, patient evaluation, and surgical and therapeutic essentials, as well as those of each of the medical-surgical specialties and internal medicine, is taught in human anatomy courses. Anatomy is not the oldest basic science for no reason, and Hippocrates (460-377 B.C.) characterized it as the beginning of medical science in terms of the nature of the body. Thus, anatomy is the earliest basic science, the historic and traditional base of medical education, and the source of the first scientific approach to sickness.

Human anatomy includes the systematic and organized study of the structural make-up of the human body, as well as an overall view of organic systems, vascular and nervous relationships, skeletal system organization, recognition of the textures of the body's tissues and organs, shapes and sizes of the viscera, and position and three-dimensional planes of the organism.

Systematic (or descriptive) anatomy, regional (topographic) anatomy, microscopic (histological) anatomy, developmental anatomy, functional anatomy, surface anatomy, endoscopic anatomy, neuroanatomy, clinical anatomy, surgical anatomy, applied anatomy, radiological anatomy, comparative anatomy, forensic anatomy, anthropological anatomy, and artistic anatomy are just some of the divisions and sub-disciplines of modern anatomy. This helps to explain why anatomy classes have so much substance and so many applications, both theoretically and practically.

In today's date, the way medical and other health sciences students study and understand anatomy needs to be examined and understood in the context of the teacher/student relationship. The many courses in the basic sciences, including anatomy, are increasingly showing signs of greater integration, a systems orientation, the encouragement of self-taught and meaningful learning, the integration of new technologies, and data purification.

At the same time, the amount of content and information on anatomy has grown, without necessarily increasing the amount of time available for it in university programs or anatomy classes. This issue leads to discussions in university pedagogical and administrative sectors about how to teach anatomy in medical and health sciences schools and faculties, which pedagogical model to apply, and what degree of requirements to set.

Human anatomy education approaches have been primarily dependent on digital/virtual technologies, and it has even been recommended that medical anatomy courses be delivered between semesters. It is critical not to undervalue the specialized teaching technique required to teach human anatomy, nor to undervalue the educational experience of students (medical and other health sciences employees in training).

The classical and regional models, systemic anatomy, and clinical anatomy have all existed, and will continue to exist, in the way that teaching and learning anatomy has evolved over time. The validity of the classical model for teaching anatomy is being questioned today, sometimes due to the obtuse concept that it is old and obsolete, but these ideas may be based on a poor interpretation of pedagogical and didactic resources (which are increasingly virtual, digital, and dynamic) with pedagogical models in the effort to transmit and develop learning in the student.

Along with other basic undergraduate courses in medicine and health sciences, anatomy provides the concepts for correlating and ordering functions, and identifying and separating the physiological findings from the pathological findings, simply establishing the basis for clinical practice by tying it directly to the clinical disciplines.

7.2. HISTORICAL CONTEXT: THE BEGINNINGS OF ANATOMY AND THE CLASSIC TEACHING MODEL

Herophilus of Chalcedon (335-280 B.C.) and Erasistratus of Ceos (300-250 B.C.) are said to have done the earliest human anatomical dissections, as well as vivisection, or live animal experiments. Herophilus was the first to discover and characterize the optic nerves and retina, as well as to distinguish nerves from tendons, demonstrating that nerves originate in the brain and spinal cord and go to the muscles.

As Herophilus' pupil and assistant, Erasistratus saw that the brain's ridges (convolutions and gyri) were more prominent in humans than in animals, and hypothesized that injuries to the fourth ventricle (rhombencephalic ventricle) caused sudden death. It's important to remember that the history of medicine (which includes anatomy and the first dissection) includes anatomical records from places other than ancient Greece, including Egypt (the Ebers and Smith papyri), India (Ayurveda), Rome, and the American aborigines.

Later, Claudius Galen (131-201 A.D.), in the ancient Greek city of Pergamum, used animal dissection methods (human cadaver dissection was prohibited at the time) to develop a wide variety of anatomical descriptions, including those of the dura mater, pia mater, corpus callosum, cerebral ventricles, pineal and pituitary glands, as well as the identification of 11 of the 12 cranial nerves.

Although he assumed that animal and human anatomy were identical, he provided functional descriptions of the structures, such as how the spinal cord controls muscles, the difference between veins and arteries (where blood, not air, circulates), how the brain controls the voice, and how the kidney produces urine.

Galen also discovered that a spinal cord injury between the first and second cervical vertebrae resulted in immediate death, and a section between the third and fourth cervical vertebrae resulted in respiratory paralysis. Associations were drawn between structure and function (functional anatomy) and between species (comparative anatomy) from the most lofty and humble beginnings of anatomy, and basic explanations of some disorders were given (pathological anatomy).

With his work "De Humani Corporis Fabrica," Andries van Wessel, also known by his Latinized name Andreas Vesalius, revolutionized medical education while he was just 29 years old. Considered to be the father of modern anatomy, he is credited with correcting some of his predecessor Galen's errors (anatomical inaccuracies of the sternum, liver, and origin of blood arteries due to Galen's concentration on the comparative anatomy model).

Although Vesalius was primarily a structural anatomist, he also supplied a few pathological references, such as aneurysms and tumors. Anatomy, which is a mix of science and art, initially satisfied a natural curiosity about the human body and who one is as a human being.

The classic model is based on the teacher-student pair, cadaver dissection by planes, description of structures, and macroscopic relationships by regions, as depicted in Rembrandt's masterpiece "The Anatomy Lesson of

Dr. Nicolaes Tulp" (1632), in which Dr.Tulp gives a class on the forearm to a group of surgeons (and one or two "interlopers" who paid to enter the "anatomy theatre").

In the nineteenth and early twentieth centuries, the greatest topical development in anatomy (descriptive and topographical) is said to have occurred, leading to the founding of the Anatomical Society of Paris (1803) by Dr. Dupuytren and Dr. Laennec, as well as the first attempt to unify anatomical nomenclature in Basel (1895).

7.3. BASIC MODELS FOR TEACHING AND LEARNING HUMAN ANATOMY

Regional anatomy, systemic anatomy, and clinical anatomy are the most prominent methodologies and paradigms for studying and learning human anatomy.

The organization of the human body by parts and segments (head, neck, trunk, and extremities), areas and regions, and defining the arrangement of the body by layers is known as regional or topographic anatomy. Surface anatomy is used to detect palpable structures (the basis of physical exploration). This model is typically used in anatomy classes in medical or health sciences schools with a dissection lab.

Systemic anatomy (anatomy by systems) is based on the study of each of the organism's systems, which explains the organism's complex and integrated functions. Clinical, medical-surgical, and other health-related professions are based on this approach. In general, systemic anatomy examines the integumentary system (dermatology and aesthetic medicine), skeletal (osteology), joint (arthrology) and muscular (myology) systems (three basic systems for orthopedics, traumatology, physiatry and physical therapy), nervous system (neurology, and, in turn, includes the sense organs; the object of study of clinical neurology, neurosurgery, psychiatry, psychology, otorhinolaryngology, ophthalmology, speech therapy and optometry), circulatory system (angiology), lymphatic system (fundamental for internal medicine and oncology), digestive system (gastroenterology), respiratory system (pulmonology, respiratory therapy and speech therapy), urinary system (urology), reproductive system (gynecology, andrology and sexology), and the endocrine system (endocrinology).

Meanwhile, clinical or applied anatomy explains and/or resolves clinical practice cases by utilizing links between structures (regional or systemic) and function. As a result, the clinical anatomy model (also known as medical-surgical anatomy) connects human anatomy to diagnosis, treatment, and surgical procedures.

This is an integrative model that encourages anatomical-clinical and pathophysiology investigation to support clinical practice. Regional and systemic anatomy, on the other hand, provide the anatomical underpinnings for applying and discovering the intriguing model of clinical anatomy. The pedagogical models of anatomy are complementary rather than exclusive, and they can be enhanced by a variety of materials and didactic tools.

Practical learning or dissection are didactic methods for operationalizing the preceding models' teachings (especially the regional and systemic models). Dr. William Hunter noted in 1770 that dissection teaches the freedom and promptness with which a live subject can be sliced or inspected. Dr. Moore stressed the importance of observation and palpation, as well as movement and dissection of various body regions, and noted that dissection is a well-established research method that can be a very engaging learning pathway if the student understands the clinical significance of the structures being dissected.

Thus, why is there a movement to discard or abolish dissection from some medical and health sciences colleges although it is the most classic and historical teaching instrument for enlivening anatomy's basic teaching and learning models? This chapter suggests a different question: why not supplement and update it with new resources and teaching methods based on imaging, informatics, and 3D and immersive models (such as mixed reality)?

The diversification of diagnostic imaging (in the 1970s and 1980s), which gave rise to radiographic anatomy, complemented cadaver dissection to combine teaching materials and enliven the learning of human anatomy. This subject allows for the regional study of deep structures and systems integration, providing information not available from cadaver studies alone, such as anatomical variations, muscle tone effects, body fluids, pressures, and organ structural detail.

However, just like clinical anatomy, excellent radiological anatomy necessitates anatomical foundations drawn from the fundamental models: regional and systemic anatomy. The complement between pedagogical models and teaching operationalization, employing diverse ways and teaching materials, is advocated once more.

7.4. CURRENT TREND IN MODELS, METHODS, AND TOOLS

Human anatomy is currently being taught and studied in medical and other health sciences colleges, and it is in a critical and transitional period. With the introduction of the digital era into an individual's day-to-day reality, over the last several years, the perception of how the morphology of the human body can be taught, learned and studied has changed, in large part spurred by the generation gap between teachers and students, as well as by administrative tensions between cost-benefit, academic quality-resource optimization and program profitability.

As a result, several medical schools have revamped their curriculums, reduced learning hours, and even reduced the details and content of their anatomy courses, while emphasizing the use of the clinical anatomy model, which focuses on the student as a future healthcare professional rather than an anatomist. The reorganization of anatomy course curricula helps to understand why anatomy pedagogy and teaching models differ throughout institutions and programs.

Modern literature expresses worry about physicians-in-anatomical training's knowledge and the impact this may have on their professional and clinical practice, as well as how poor anatomy instruction might undermine patient safety across numerous healthcare systems.

Dissection, prosecution, plastination, anatomical informatics, imaging, and living anatomy, as well as other teaching models based on readings, integrated curricula, and systems-based curricula, are all mentioned as modern tools and methods for teaching anatomy in Estai and Bunt's 2016 critical review of the literature.

Individual or small group experiences in which anatomy is actively investigated in bodies (cadavers), sectioning in each of the body planes to divide and discover the anatomical structures being studied, are known as dissection. This didactic tool, together with lectures and master courses, has been the primary teaching tool in educating anatomy utilizing the regional and systemic paradigm for more than 400 years.

Some critics believe it is quite an expensive method that requires a significant time investment and is outmoded. It has been superseded in certain medical schools by approaches such as prosection mixed with other teaching modalities.

The anatomical study of another live being without the need for dissection is known as living anatomy. For example, identifying forearm tendons or palpable bone portions on the extremities through a physical examination of their own bodies, peers' bodies, or simulated patients. It entails painting or drawing anatomical structures on the skin in order to comprehend the underlying morphology, as well as providing valuable spatial correlations. It also contains specially made apparel to help pupils comprehend the notion of dermatomes, for example.

The use of cadavers, prosection, and plastination, as well as plastic replicas, in combination with master courses, is the foundation for education at various universities.

A research field has emerged from the comparison of anatomical models, methodologies, and instruments. For hundreds of years, corpse-based education and learning has been the primary teaching method, however in light of the new modalities stated, there are various views on whether thorough cadaver dissection is still appropriate for current university study.

There are studies that support the utility and usage of cadavers in the decade between 2010 and 2020, in which various universities in the United States, Australia, and New Zealand, which had abandoned cadaver-based anatomy teaching a few years before, restarted it a few years later.

7.5. MODERN TRENDS IN CLINICAL ANATOMY TEACHING

The paradigm of medical education has shifted as a result of recent rapid breakthroughs in information and communication technology. Computer networks, web-based learning, and portable network devices are becoming an increasingly important part of medical education's learning environment, influencing curriculum design and reshaping the medical curriculum.

Students have access to a great amount of material; automated learning systems and interactive programs are all available for use in class and for personal study. These new information and communication technologies are being employed to improve the teaching and learning environment. They have the ability to help students learn and solve problems, allowing for better integration of the didactic approach with clinical practice and a smoother transition between classroom learning and clinical practice.

Formal lectures are still the backbone of medical education and the most common method of learning, as they allow for the delivery of vast amounts of factual knowledge to a large group of students in an orderly manner. Using techniques such as drawing on smart boards, video snippets, "moving" graphics, and others, current teaching technologies enable more engaging lectures.

Students become active participants in the learning process when clinical scenarios and question discussion are used during lecture time, especially when feedback technologies such as clicker technology are used. Pure lecture classes may be increasingly out of touch with how students engage in their world, as modern students are predominantly active learners. Relatively 'passive' lectures will, of course only get one so far, and apprenticeship supported by self-directed study by individual teaching still remain necessary for learning specialized and high-level skills.

Students now use e-books on laptops or tablet devices instead of traditional textbooks. In fact, many medical schools no longer use dissection in fundamental courses like first-year anatomy at all; instead, they use expertly dissected plastinated specimens and virtual 3D body systems. Online access to the class schedule, professor presentations, and tests is now accessible.

The majority of courses are provided online using technologies like the Learning Management System (LMS) or a Virtual Learning Environment (VLE). Effective course management systems manage teacher-student contact while also handling and controlling the delivery of learning courses. The LMS keeps track of the learners' actions, progress, and compliance with course assignments, as well as facilitates peer-to-peer and professor-to-student contact.

Many colleges offer lecture broadcasting to students, allowing them to attend lectures from anywhere at any time in the 'virtual classroom.' Information technologies are especially significant in today's medical curriculum because the shorter anatomy course frequently compels the teaching of all anatomical elements of the subject away from its clinical application.

Medical students, on the other hand, must think like future doctors from the start. This is where technological innovations are critical for bridging the gap between pure didactic lectures and interactive learning, bringing clinical scenarios closer to life by enabling on-the-spot problem-solving. The following are some examples of these advances. Real-life situations are simulated or imitated utilizing technology.

Particularly useful are anatomical locations or clinical procedures. A virtual dissecting table is an example of simulation. It allows users to see the 3D anatomical details of a virtual cadaver in a realistic way. CT scan data was used to create a full-body gross anatomy model. Any data from CT, MRI, and ultra-sound scanners can also be opened with the Table.

The students can redo and undo the dissection as many times as they like, unlike cadavers. By virtually slicing, layering, and segmenting the anatomy, multiple body perspectives can be customized, allowing for natural and intuitive operation methods.

For trauma, skeletal fractures, and joint dislocations, traditional instruments such as plain radiography, as well as chest and skull x-rays, are often used. In most anatomy courses, computed tomography (CT) and magnetic resonance imaging (MRI) are both heavily used.

Ultrasonography is frequently employed in all fields of medicine, and ultrasound images, along with radiographs, CT scans, and MRIs, should be available in Gross Anatomy labs. Because ultrasound is a non-invasive imaging modality, a machine could be set up right in the lab to give students the experience and abilities they need.

Ultrasound technological advancements (probe size and frequency range) enable the scanning of many body parts. Endoscopy (gastroduodenoscopy, colonoscopy, and bronchoscopy) is a non-surgical treatment that allows doctors to look within empty organ cavities. On the radiography station, endoscopic images of the esophagus, stomach, duodenum, large intestine, and bifurcation of the trachea may be used to strengthen students' awareness of various clinical applications presented during the lab session.

There is a pressing need for learning aids that offer representations of the real human body in three dimensions for those who must learn or relearn human anatomy without access to cadavers. In the presentation of three-dimensional images, there are two methods: stereoscopic vision and rotation. Stereoscopic vision is the foundation for all 3D visuals and movies, and it is employed for everyday depth perception. All of these visual aids necessitate specialized equipment (3D glasses and screens).

There is a third option: rotating the object. Object rotation is something individuals perform instinctively whenever they want to understand more about a solid object. It sends a series of images to one's seeing brain, which it combines to form a three-dimensional mental representation. Any two-dimensional representation of the item that is seen afterward makes sense in three dimensions once one knows that three-dimensional image.

Many applications for portable handheld electronic devices increase access to resources regardless of location. They are frequently used by students for medical inquiries, interactive games, and to familiarize them with electronic tools that will be useful for patient management and treatment decisions in the future.

Various medical applications are dedicated to anatomy and physiology and other related topics, while others focus on medical problem solving, diagnosis, and treatment. Furthermore, social media platforms such as Facebook, Twitter, and other apps allow students to interact and organize tutorials, group study sessions, information sharing and peer-to-peer learning.

Any course must include some form of assessment. It's a final output of the teacher's work, and an assessment of the students' knowledge. The majority of quizzes and exams are currently computer-based and incorporated into the Learning Management System. Starting with weekly tests and ending with the NBME subject shelf, which is regularly utilized by students for medical questions, interactive activities, and allowing students to become familiar with electronic apps that will later be beneficial for in-patient management, and treatment decisions.

Furthermore, LMS also provides surveying capabilities. Pre- and post-course evaluations can both be done through surveys. They enable the instructor to assess students' knowledge of topics in order to aid in course design or to make various aspects of his teaching easier. Surveys can be used as a post-course tool to assess student progress or instructor performance.

7.6. EVOLVING TRENDS IN ANATOMY- A GLOBAL PERSPECTIVE

Anatomy education has progressed from a simple chalk and board approach to one that includes 3D models, digital dissection tables, PowerPoint presentations, and role plays. In today's world, a teacher must continue to update his or her knowledge in medical education by taking advanced courses. For all of us, webinars have become the new normal. They're a good and safe way to meet and share thoughts about current anatomy trends and new perspectives. Students are facilitated and mentored by modern professors, who do more than just teach.

Students have developed as well, and they no longer rely on books to understand anatomy. To keep their skills up to date, they use YouTube, numerous applications, and hands-on workshops. New additions to the anatomy classroom include simulation labs, a C-arm, and an ultrasound equipment. Endoscopic anatomy is the most recent trend, which teaches students more than just anatomical positions and gives them a whole new way of looking at the body. Medical ethics, which was never officially taught in anatomy classes, is now becoming more popular as part of competency-based medical education (CBME curriculum).

7.6.1. Virtual Reality

Virtual reality, augmented reality, and mixed reality are rapidly growing technologies that are being used in the classroom to take students on virtual field trips of the human body.

- Pros: Appeals to visual learners.

  Cons: Students lack the personal human touch.

Figure 7.1. *Virtual reality.*

Source: Image by Wikimedia

7.6.2. Digital Readers/Tablets/ Mobile Technology

Bulky hard-copy textbooks are increasingly being replaced by digital textbooks that may be accessed via a tablet or mobile phone. Rather than prohibiting the use of mobile devices in the classroom, some schools are incorporating technology into the learning process through instructional apps.

- Pros: The cost of acquiring new textbook versions every few years is eliminated because digital content is updated on a regular basis. These instructional apps can be tailored to your preferences.
- Cons: To fully execute, schools would need to equip each student with a tablet and set up a procedure for dealing with lost, damaged, or stolen property.

7.6.3. 3D Printing

It is an excellent application for creating organs and models. For an interactive and learning experience in a group, one can create models for students.

- Pros: 3D printing can be useful for both kinesthetic and visual learners can decrease the time spent on planning and designing models.
- Cons: For a student, 3D printing technology is an easy way out in place of their own physical model creation.

Figure 7.2. Working on a 3D printer.

Source: Image by PxHere

7.6.4. Gamification

Quiz/crossword puzzles/snowball/fishbowl technique/ role plays for AETCOM modules

When having fun, students tend to learn better. Gaming can be used in the classroom by joining the fun part of play with conceptual learning.

- Pros: Students get enthusiastic because it increases engagement. Immediate feedback can be received.
- Cons: The fun game is not effective at teaching all concepts. One has to devise effective use of games for learning programs.

Figure 7.3. There are many gamification techniques used to teach anatomy.

Source: Image by Wikimedia

7.6.5. Cloud Technology

Flipped classroom and self-directed learning

Digital textbooks, courses, videos, and assignments are all stored and shared in the cloud. It allows students to engage in live conversation. This technology allows students to benefit from a new teaching approach known as "flipped classrooms." Students in this technique watch a lecture before class and then discuss it in class. In the classroom, students can engage in group work and analytical exercises.

- Pros: It allows students to access information from any internet-connected device and to communicate with their teachers quickly through live chat options.
- Cons: Every student should have decent internet connectivity. Hacking and other security vulnerabilities might be a source of concern.

Figure 7.4. Cloud technology promotes self-learning.

Source: Image by PxHere

7.6.6. Magic Mirror

The Magic Mirror is a user interface approach that imitates a standard mirror but adds nonphysical visual input to the optical effect. The user's image is captured with a camera and displayed on a screen that resembles a mirror.

- Pros: It overlays medical data on the user's body and displays additional 2D and 3D information based on the user's requirements. The concept of medical education and rehabilitation is expanded by this approach.
- Cons: Costing and unnecessary usage may be of grave concern.

7.6.7. Artificial Intelligence (AI)

In the educational field, AI is utilized to automate grading and feedback. It allows students to learn at their own pace. It gives an individual a better understanding of how a pupil learns.

- Pros: Requires fewer human resources. Saves time for teachers by grading and providing feedback on their behalf.
- Cons: It lacks a personalized human touch. There's no real-life feel.

Figure 7.5. AI aids learners to learn at their own pace.

Source: Image by pixabay

7.6.8. Simulation

Advanced manikins are employed in simulated patient situations in this instructional technology. It's also known as a high-fidelity simulator or a human patient simulator.

- Pros: This technology aids in the control of full-body manikins by computerized systems. These are set up to react realistically to the actions of the students. The manikins can talk and breathe, as well as sleep and wake up. Structures, functions, symptoms, and medical skills are best presented with these features.
- Cons: Costly and difficult to maintain.

7.7. EDUCATION IN THE DIGITAL AGE: TECHNOLOGICAL TRENDS IN ANATOMY EDUCATION

7.7.1. Technology-Based Learning Environments

The promise of this new information technology was regarded as a lever to improve the quality of education with the rise of the microcomputer in the 1980s, and this development has prompted high hopes (De Corte, 1994). Rapid advancements in digital technology have resulted in the development of new instructional methodologies and platforms.

Many tools developed during the digital era are now utilized in the teaching of anatomy. The debate over these technologies, on the other hand, is beyond the purview of this research. However, several digital technologies that are considered to be vital, including personal computers, the internet, and mobile smart devices, have also contributed to anatomy instruction.

Massive-open-online-courses (MOOCs), infographics, digital 3D simulations, augmented reality (AR), virtual reality (VR), and wearable technologies are some other digital technologies that arose in addition to these developments.

7.7.2. Computers and Computer-Based Learning in Anatomy Education

With the widespread use of personal computers, computer-assisted education has long been a priority in the field of education. Not just in affluent countries, but also in emerging and undeveloped countries, computers have emerged as effective and realistic educational instruments (Kozma & Anderson, 2002). Anatomy is a crucial course in medical and health sciences education, and it is through this course that physicians begin to build their clinical abilities.

Even though human dissection education was not emphasized in the mid-nineteenth century, it was feasible to get a doctor's degree diploma in the United States. Many medical educators, on the other hand, believe that disciplined medical students should serve as a practical foundation for future surgical procedures in general practice. Here, it is important to note that as a result of this awareness, dissection is becoming more popular among medical students (Warner & Lawrence, 2006).

Anatomy must be well understood for safe clinical activities in general, and surgical specialties in particular (Turney, 2007). The use of cadavers is the most significant aspect of anatomy instruction, however, due to the scarcity of human cadavers, the use of anatomy models created using specific digital software is becoming more popular (Hoyek et al., 2014).

Because the anatomy learning process is dependent on diagrams and anatomical images, in particular, using computers for anatomy teaching is particularly convenient. As a result, if the course is taught using multimedia tools on a computer, the student will be more engaged (Ur-Rehman et al., 2012). Before presenting lectures to students, computers allow the instructor to develop a strong organizational structure in the appropriate form (Cassady, 1998).

Students and instructors can benefit from computers in terms of both time and money. At the same time, computers allow pupils to see magnetic resonance (MR) and x-ray pictures simultaneously (Ur-Rehman et al., 2012). If it is well-designed and the curriculum is well-integrated, computer-assisted learning can be highly effective in anatomy training (Tam et al., 2009).

7.7.3. Internet-Based Learning in Anatomy Education

Due to an increase in the number of students attending anatomy courses and required application hours, there are certain challenges in anatomy training. Small groups of students cannot be taught anatomy lessons due to the limited utilisation of laboratories and the limited quantity of teaching professionals (Ozer et al., 2017). Educators, on the other hand, are currently working on new approaches.

The advancement of video streaming and broadcasting technology, as well as the development of web-based computer-assisted learning platforms like Moodle and Blackboard, as well as the ability to combine various communication and information technologies, have made it easier to integrate students into the learning process.

Web-based visual courses are useful in teaching anatomy using internet-based learning (IBL) in general because they create solutions that are personalized to each student's intellectual and psychological profiles and needs. In addition, IBL can help students succeed, especially in big groups (Ozer et al., 2017). One of the most important aspects of effective and long-term learning in medical education is the availability of e-learning materials via learning management systems (LMS) (Gray & Tobin, 2010).

7.7.4. Mobile Devices and Mobile-Based Learning in Anatomy Education

Tablet computers soon found application fields in numerous nations after their introduction. Since the popularity of mobile devices has grown, many scientists are interested in incorporating these technologies into education and the classroom, believing that kids will benefit from them. As a result, anatomy classes are now taught using tablet computers.

For example, Lynn et al. (2015) found that employing the anatomy of muscle and skeletal systems loaded on iPad computers improved students' learning experience, boosted their cooperation with their classmates, and helped them master course materials. Smartphones are the most widely used mobile devices, and they are growing in popularity in both personal and business settings.

We have seen the advent of new technology that has revolutionized many parts of our society, commerce, communications, and education during the previous decade. Smartphones have become widely accepted in many nations, allowing information to be accessed in ways that were previously impossible (Gavali et al., 2017). For a variety of reasons, smartphones are viewed as an appropriate instrument for imparting further education, particularly in underdeveloped countries.

This is widely known to be due to the fact that mobile phones are more widespread and have a higher penetration rate in developing countries. Nonetheless, the expanding impact of mobile phones on the educational environment in poorer countries requires further research (Walk et al., 2010). Smartphones are now employed in a variety of medical disciplines. Furthermore, the number of applications and programs created for medical usage is growing by the day.

The smartphone, as an educational instrument, has the ability to teach both theory and application of the needed information. According to several studies, smartphones can be an effective instructional tool for medical students (Robinson et al., 2013; Jamali et al., 2015; Wenting et al., 2017; Mackay, Anderson & Harding, 2017; Risling, 2017).

7.7.5. MOOCs (Massive Open Online Courses) In Anatomy Education

MOOCs are massive open online courses in which thousands of people can participate at the same time. MOOCs are open online courses established by higher education institutions throughout the world and made available to the public through commercial platforms such as FutureLearn, EdX, and Coursera. MOOCs have been chastised for the video classes' poor pedagogical value, with merely multiple-choice exam questions.

When compared to open educational materials, it is believed that MOOCs are now unable to integrate the curriculum and provide a real benefit over them (Doherty et al., 2015). Despite the fact that there is numerous research articles on medical education via MOOCs in the literature, the potential role of MOOCs has yet to be fully explored (Pickering et al., 2017).

7.7.6. Infographics in Anatomy Education

Visualization enables students to get complicated information in the learning process in a more efficient and effective manner. Infographics, a visual media, are becoming more popular as a way of communication. As humans, we are more interested in pictures, and aesthetically produced visual expressions and storylines easily entice us.

People are undeniably assaulted with more information on a daily basis than they were 20-30 years ago (Evans, 2016). Every day, we are bombarded with new information from a variety of sources, including new circumstances, social media tools, radio and television, e-mail, ads, billboards, daily office work, and a variety of other sources.

In actuality, our ability to process large amounts of data is restricted. As a result, infographics featuring images, captions, and data catch our attention more effectively and simply than other types of information attacks. Studies on how to utilize the benefits of infographics in education have been examined based on these beneficial features (Ozdamli & Ozdal, 2018).

Infographics make the process of conveying and comprehending complex and massive volumes of data much easier. Infographics, for example, highlight crucial texts and data that a reader would overlook in a long text. In this regard, not only in social media, but also in academic forums, the use of infographics is fast expanding. Ozdamli et al. (2016) studied the students' opinions and attitudes on the anatomy course that was provided to them using graphical materials.

More than half of the students (58.6%) thought that taking an anatomy course with infographics will help them learn the subject more effectively. More importantly, 86.4 percent of students said that using infographics helped them understand the lessons better.

7.7.7. Three-Dimensional Digital Simulations/Programs in Anatomy Education

In-class lectures, textbooks, atlases, and, if accessible, cadaveric dissection are all examples of traditional anatomy teaching methods. Cadaveric dissection is an important part of anatomy teaching. Cadaveric dissection is supposed to improve pupils' manual dexterity and communication skills (Aziz et al., 2002).

The reality is that corpse tissue is not as authentic as living tissue, which makes cadaver dissection difficult. Because the anatomical architecture of cadavers is twisted, cadaver tissue differs significantly from that of living human beings.

Indeed, the tissues of cadavers and living human bodies differ significantly in terms of quality, color, and textural anomalies. Another key issue is that, with the exception of surgical branches, the value of cadaveric dissection in anatomy lectures for practicing students is debatable. This is due to the fact that some students (such as those studying pharmacy, audiology, nursing, and other related fields) do not require cadaver dissection knowledge.

Other drawbacks of corpse use in anatomy instruction include challenges with cadaver procurement, transfer, preservation, high expenses, short-term usability (since tissues are disassembled and fragmented following dissections), and psychological stress. Because of these drawbacks, the efficiency of cadaveric dissection in anatomy education has been questioned.

Alternative techniques (including digital technologies) for anatomy instruction have been proposed for these reasons, particularly since the introduction of computer technology. Anatomy instruction can be aided by a variety of computer tools. Popular examples include digital anatomical atlases with 2-dimensional materials and multimedia versions.

Virtual anatomy (VA) systems, on the other hand, have grown in popularity in recent years because they supply students with 3-D anatomical images and materials. VA permits anatomical structures to be observed from any angle in anatomy teaching, and VA is not limited in terms of image angles. In the same picture, VA can display the morphology of anatomical structures as three-dimensional images, such as their positions and spatial relationships (e.g., connections to other organs, vascularization, and innervation). As a result, when utilized in conjunction with certain learning goals, interactive 3-dimensional (3D) visual resources have a huge potential to replace traditional anatomy education techniques (Brenton et al., 2007).

Students stated that interactive systems like VA were beneficial to them, and that they had made notable developments in understanding the spatial relationships of organs (Preim & Saalfeld, 2018). Students demonstrated that 3D anatomy models were superior to 2D ones in another study. The predominant trend in anatomy teaching, according to the same study, is toward 3D digital models (Azer & Azer, 2016).

7.8. AN INTERACTIVE VR SYSTEM FOR ANATOMY TRAINING

In a virtual environment, several training apps are set up. Users may be exposed to catastrophic scenarios that directly influence their security and lives in these simulated environments. In battle strategy training and surgical surgeries, this scenario is very prevalent. As a result, medical professionals highly advise using simulators and virtual reality technologies to train future doctors in surgery.

This solution improves training efficiency. Realistic virtual environments have been created in order to immerse the user in the environment and allow him to imitate real-life scenarios. Today's medicine is well-versed in technical and technological advancements. As a result of this evolution, the trainer's job becomes increasingly difficult to do and explain to the students.

In recent decades, virtual reality (VR) technologies have provided students and health professionals with a fantastic opportunity in medical education. In addition, the popularization of new technology pushes medical students to become acquainted with virtual reality.

7.8.1. Contributions of Virtual Reality to Medical Training

Due to the vast volume of data to be given to learners, traditional training approaches do not always suit modern educational objectives. At this moment, virtual reality (VR) may be a viable option for the data management issues that have plagued prior methods.

Virtual reality (VR) provides fresh solutions to all control, simulation, and communication issues. It is portrayed as a step forward from traditional simulation methodologies. Virtual reality training has a number of advantages over traditional training, including: It allows us to complete tasks without risk: when we work in immersion, we are in contact with virtual objects that do not pose a threat to our safety because we are manipulating multiple objects (objects of an operating room such as the scalpel).

Tolerance means that we are allowed to make and correct mistakes without fear of security being jeopardized because the errors are formative (ex: to make errors on a virtual body no risks if one made these errors on a human body who can put his life at risk) In the same way, virtual reality allows us to create scenarios with realistic sensations in order to immerse the learner in more realistic situations. We can simulate more realistic environments and in conditions that would be impossible to achieve in real life.

Another advantage of VR is its accessibility, as it allows you to receive training without regard to time or location. Training in VR does not require a specific time or attendance requirement in a room. You can take your course at any time and in any location. In addition to the benefits mentioned above, the cost and occupied space are reduced because VR takes up less space than a model or a traditional skeleton.

Furthermore, the same equipment can be used for other modules and even other training, making VR training less expensive than traditional teaching methods. Immersion in a virtual world enhances learning and the environment by incorporating crucial sensory qualities in a variety of settings.

7.8.2. Related Work

There has been a lot of work accomplished in the profession of teaching with the aid of VR since the latter has attracted the interest of researchers in order to simplify the task for trainers and aid future doctors in the assimilation of the course, so there are those who have basically made a state of the art on both teaching techniques and drawn conclusions on the benefits and drawbacks of each, while others have had to propose applications in various medicinal specialties.

Khwanngern et al. researched a rare case of craniofacial disorder, which was viewed as a rare case of this phenomenon that student doctors may not have a real case of the cutting process and make that in theory, which is not helpful for students. To help resolve this, they developed an application that simulates a human skull in an operating room and can cut, drill, and manipulate the latter thanks to motion controllers.

The procedure of cutting the mandibular bone, which is a key action during jaw surgery, is the topic of this study (lower jaw). The cut needs extreme precision, and even a minor error might injure the face nerve, resulting in facial muscle paralysis. The proposed concept is intriguing, but it does not simulate reality because it simulates the skeleton of a skull, whereas in reality, we are confronted with a human body, which we would prefer to simulate and proceed to cut, and second, we visualize the helmet controllers, which may obstruct us during the cut due to a loss of precision.

Alfalah et al. conducted a comparison study between traditional medical education methods and virtual reality (VR) technology. The latter is a tool that provides additional means to guide in aims to enhance the quality of skills and meet the requirements of modern medical training in order to overcome the difficulties faced by students and teachers in conveying the message. A comparison of the two teaching approaches in VR and the traditional method was undertaken.

The purpose of the experiment was to test students by giving them a questionnaire on both teaching techniques, and the VR offers the ability to operate the organ (human heart) in 3D. Dissect the 3D cardiac model in layers to clarify anatomical relations of different sections, study the information on each component of the model, and investigate the characteristics offered in the system, which do not exist in traditional methods.

Huang et al. did a study on student adoption of new technologies like virtual reality. After setting up a learning application in VR and delivering a questionnaire to students while they were using it to get their feedback, we calculated that the percentage of acceptability is high. Further, these programs allow students to take their courses without being bound by time or face-to-face limits; we can follow the course of our choice at any time and in any location.

De Faria et al. believe that the methods used in the formation and teaching of future doctors are extremely complex, especially those that are based on conferences or laboratory dissertations. To do this, they formed three working groups with different tasks and compared their perceptions during different attempts using the two learning methods and comparing the statistics of these, the results reveal satisfactory results using virtual reality in the teaching that the students better assimilate the information.

Izard et al. have outstanding 3D human body anatomical models, as well as numerous VR applications based on Toshiba Medical Systems' DICOM and Asteion CT scans at Complejo Hospitalario. Universitario de Salamanca, using the skull study protocol: one in anteroposterior projection and the other in a lateral position. By performing research of the potential and contribution teaching of VR in education, users can navigate between different regions of the human body utilizing the stereoscopic system and interact to make a choice about which information to display.

Mathur researched a very essential and significant case by starting each project that the cost, and most of the existing virtual reality systems are really expensive, particularly when it comes to specialized systems, so to fix it, he proposed a system with an oculus helmet and razer controllers called hydra.

7.8.3. Conceptual Model

In this case, an application was created by providing interactive educational tools that allow users to interact with the application in order to make decisions about which explanations to display in the virtual environment. The entire experience is based on 3D immersion, and we provide interactive educational tools. We have an avatar of a human body in the surroundings, or the user can move this body with the touch of the HMD helmet.

We offer users a menu that they can organize using touch or the latter can perform on the rotation of the human body and on the transparencies in order to separate the skeletal structure alone or just the organs; by aiming an organ or a bone, you will have the latter's concept appear on a side panel, and thanks to users can use the laser keys to scroll through the text and read it completely.

The effort entails the creation of an ogrVR application, which is a teaching platform (application) aimed at medical students, with the advantage that it may also be used by others who are intrigued about human anatomy and want to learn more about it.

The desktop application orgVR is an instructional application on human anatomy which allows the user wearing an oculus rift helmet and using touch controllers to engage with the human body. The possible interactions are multiple (X-ray, rotation, change of opacity, appearance of documentation, and so on) and are implemented using c # and xml scripts and shaders that we created on organs that were modeled.

The suggested environment is an immersive and engaging environment, in which the user is equipped with an oculus rift helmet and touch, or the latter is immersed in a lab and can use the interaction concepts that have been developed due to the touch (navigation, manipulation and selection).

Navigation: is the capacity to move in a virtual environment and make translations when being immersed, the proposed environment is an interactive environment that opens up the possibility of navigating in this environment thanks to the transcriptions and the motion sensor which translate among our condition in the physical and the virtual world.

Selection: the concept of selection is utilized when you want the description of an organ or other thanks to the laser of the oculus rift touch it is quite enough to target (target) this organ or the body part you would have the description of the latter which appears likely in pop-up window format on the side and then users can scroll in order to be able to read the full text, thereby enabling to choose the various menu items.

Manipulation: The application has a menu with numerous capabilities that allow you to modify the human body, including transparency and rotation.

The orgVR usage scenario:

After starting the program, the first step is to put on the oculus rift helmet, attach the oculus touch controllers, and connect the two movement sensors to detect our position. The user is immersed in the first scene or window, where he has the option of manipulating or studying the gender of the body (male or female).

Users will either be transferred to another scene or immersed with the body they have picked for study once they have made their decision. Manipulate and interact with this body using the side menu, such as rotating it in order to rotate it on itself, and toying with the different levels of opacity. If user wish to have a description or the role of the latter, they simply hit the button of the keys with their hands, and a window in pop-up format opens, which they can scroll to read the complete message.

If user want to study an organ in greater detail, simply aim the laser at it and push the trigger; user will be transferred to another room (scene) or isolated with the selected organ. The availability of a side menu, similar to previous scene, enables them to rotate the organ or cut it, move the cutting plane, so that they can catch this organ with the virtual hands present thanks to the touch, play on opacity, which implies transparency, so that they can also reload the scene to begin over or return to first scene in which the human body is present.

7.9. PREPARATION AND INTEGRATION MODELS

This is the most time-consuming and difficult element of the project; it entails the development of assets. Researchers were dissatisfied with the assets available on the internet and wanted to create their own graphics that were more realistic and detailed. They had to employ a variety of modeling tools to accomplish this.

7.9.1. Modeling (Sculpting)

Because the goal was to improve the realism of the models and user immersion, we used the Blender software to produce realistic models. The highpoly models were created (large number of polygons).

7.9.2. Rotopology

Highpoly models should not be used in 3D scenes because the large number of polygons has a negative impact on performance. This is why 3D modellers use a technique called retopology, which uses an existing 3D model to redo (improve) its topology. This will entail acquiring the ability to magnetize existing surfaces, as well as new points, edges, and faces. As a result, the extrusions and transformations of the new topology will properly follow the faces of the model object, and the model will have a lesser number of polygons after applying this technique, which will increase performance without losing model features. For this phase, they used instant mesh.

7.9.3. UV Mapping andUV Unwrapping

A UV map is a flat representation of a 3D model's surface that is used to wrap textures conveniently. UV unpacking is the process of generating a UV card.

After you've produced the polygonal mesh, you'll need to "decompress" it in a UV map. Now it's time to bring the mesh to life and give it a more realistic appearance. However, because textures are always based on a 2D image, there is no 3D texture. UV mapping is the process of transforming a 3D geometry into 2D data so that a 2D texture may be wrapped around it.

7.9.4. Texturing and Painting

In this final stage, they used Substance Painter to apply a texture to their 3D models. A texture is an image that represents a surface and can be used to simulate its appearance when applied to a 3D object. Textures are commonly employed in video games, and they provide an aspect that is near to realism. The models are now ready to be exported to Unity after this phase.

7.10. RESULTS AND TEST AFTER USING VRIN ANATOMY

Following the creation of assets, the task of integrating them into their virtual environment arose. The creation of interaction mechanisms in order to provide better visibility and perception to the user is the biggest hurdle after asset integration. They created and employed numerous techniques of interaction with objects in our virtual environment, including shader methods or raycasting, to enhance the functionality of our environment for better, more detailed interaction because shader functionality allows us to execute a variety of tasks (cuts, grabbing, etc.).

7.10.1. Pointing Laser and Interaction

The most difficult problem they ran into was implementing pointing on the organs even while managing interaction with the UI menus. This prompted us to take keen attention in Unity's ray jets (raycasting) system. In a 3D scene in virtual reality, the user cannot use the mouse, so we instated a laser to act as a pointer, enabling interactions with the virtual environment. The upshot of applying the two raycasting scripts, as shown in the following diagram, is that our laser is functioning.

7.10.2. Shader Managing the Object Cut

The cup was the most important interaction we wanted to achieve, but it was also the most difficult. Here it is explained how the researchers went about achieving it.

First and foremost, they needed to define how researchers wanted the user to perform the cut from a methodical standpoint (by what he will apply the cut). They decided that it would be more realistic and interactive if the user could perform the cut by passing a glass plane through the body, which can be manipulated either by catching it directly with the hand or by using sliders on the UI menu.

The cutting effect is achieved by applying a "OnePlaneBSP" shader to the organ; this shader calculates the position of the organ's vertices by adding them to the cutting plane's position in order to identify the vertexes of the organs that are above the plan and prevent them from being drawn on the screen.

7.10.3. Shader Managing Opacity and X-Ray Effect

The opacity change and the X-ray impact are both entirely visual effects that we created with the shader "Transparent Diffuse ZWrite," a simple transparent shader, and the script "Opacity Controller," which delivers the effect to the organ. The second effect is created using a simple "Xray Effect" shader, which is essentially a transparency shader that applies a white color to all vertices to create an X-ray effect.

7.10.4. Object Grabbing Implementation

We'll go through how they implemented grasping items in the final section. Using Unity's oculus integration kit, they attached hands to the user avatar and a simple collision circle all-around objects made it feasible to grab objects easily, just as in the real world.

7.11. CHALLENGES & OPPORTUNITIES IN ANATOMY TEACHING

At this point, it's time to think about the triumphs and failures of anatomy curricula, as well as how to teach the concepts of anatomy to tomorrow's doctors so that they might improve their clinical skills. Is the dissecting room a remnant of the past or a source of inspiration for the future?

- Anatomy is believed to be an exhausted science, and this is one of the dissident challenges that must be addressed.
- Learning through dissection is becoming less popular.
- The contents of anatomy classes are not well-defined.
- Maintaining cadaver-based education is risky due to the insecurity of cadaveric supplies.

This scarcity emphasizes the importance of utilizing modern technology and communications. The methods of teaching anatomy and the objectives of studying anatomy are incompatible. Excessive information precludes students from distinguishing between what is important and what is optional in clinical practice in many departments, where lecture-based training and several hours of dissection are the norm.

Finally, anatomists' professional status has been diminished. Because of their non-medical backgrounds and advancement expectations, anatomy teachers may place a greater emphasis on research than on teaching.

Medical practitioners are currently having difficulty identifying structures, analyzing images, determining surgical approach methods, and deciding on possible outcomes due to a reduction in anatomy teaching. About a third of resident physicians in the United States are not well educated in anatomy, which suggests that some medical errors must occur33 due to a lack of anatomical knowledge.

Technology should be included into the curriculum, but not to the point that it completely replaces cadaver-based teaching, which should occur early in the program and be clinically oriented, as it is a critical component of medical education and training. The following suggestions are intended to keep cadaveric material available to our students.

This enables an early, active learning experience with death, in which facts are confirmed from primary sources, discoveries are appreciated and compared to others' interpretations, and a three-dimensional understanding of the body, its variety, and pathology is developed. Students' manual dexterity, teamwork, and communication skills will all increase as a result of this procedure. All of this will boost their confidence and broaden their clinical knowledge.

The College of Medicine & Health Sciences at Sultan Qaboos University, as a regional center for the new Intercollegiate Membership of the Royal Colleges of Surgeons UK examinations (Edinburgh), believes that a thorough understanding of anatomy is essential and mandatory as a prerequisite to higher surgical training. Anatomy education in surgical specialties has to be enhanced.

Is the dissecting room still relevant in the education of our undergraduate and graduate students? Our answer is a qualified yes: a thorough understanding of anatomy is required to effectively define and treat a patient's ailment. Dissection is still the most effective technique to learn anatomy as a dynamic foundation for problem-solving.

7.12. CONCLUSION

It is really important to have a sound knowledge of human anatomy, especially for being a good physician or a practicing surgeon. In spite of the digitalization and modernization of the education sector, traditional cadaveric dissection still occupies the leadership of anatomy education. In the modern world, it is becoming an untold truth that no modern technologies such as virtual dissection or 3D model printing could replace cadaveric dissection in its way of teaching students. Recent research studies conducted on modern anatomical teaching methods for medical and non-medical students and their influences are being regularly reviewed so as to enhance the teaching and learning opportunities. Despite new technologies, traditional cadaveric dissection helps students to acquire more practical knowledge and still occupies a prior position on its way.

REFERENCES

1. Aouam, D., Zenati-Henda, N., Benbelkacem, S. and Hamitouche, C., 2020. An Interactive VR System for Anatomy Training. Mixed Reality and Three-Dimensional Computer Graphics, [online] Available at: <https://www.intechopen.com/chapters/71251> [Accessed 22 June 2022].
2. Mavrych, V., 2016. Modern Trends in Clinical Anatomy Teaching. MOJ Anatomy & Physiology, [online] 2(1). Available at: <http://medcraveonline.com/MOJAP/MOJAP-02-00035.pdf> [Accessed 22 June 2022].
3. Sahin, T., 2019. Dijital Çağda Eğitim: Anatomi Eğitiminde Teknolojik Eğilimler. Uluslararasi Kibris Universitesi Fen-Edebiyat Fakultesi, [online] Available at: <https://dergipark.org.tr/en/download/article-file/804828> [Accessed 22 June 2022].
4. Sharma, A. and Kumar, A., 2021. Evolving trends in anatomy- A global perspective. Indian Journal of Clinical Anatomy and Physiology, [online] 8(3), pp.159-161. Available at: <https://www.researchgate.net/publication/355286303_Evolving_trends_in_anatomy-_A_global_perspective> [Accessed 22 June 2022].
5. Suárez-Escudero, J., Posada-Jurado, M., Bedoya-Muñoz, L., Urbina-Sánchez, A., Ferreira Morales, J. and Bohórquez-Gutiérrez, C., 2020. Enseñar y aprender anatomía. Acta Médica Colombiana, [online] 45(4). Available at: <http://www.scielo.org.co/pdf/amc/v45n4/0120-2448-amc-45-04-48.pdf> [Accessed 22 June 2022].

INDEX

K

L

M

N

O

P

Q

R

S